FIRST Robots:
Rack 'n' Roll

ROCKPORT

First published in the United States of America by
Rockport Publishers, a member of
Quayside Publishing Group
100 Cummings Center
Suite 406-L
Beverly, Massachusetts 01915-6101
Telephone: (978) 282-9590
Fax: (978) 283-2742
www.rockpub.com

Library of Congress Cataloging-in-Publication Data

Wilczynski, Vince.
    FIRST robots : rack 'n' roll : behind the design, 30 profiles of award-winning robot designs / Vince Wilczynski, Stephanie Slezycki ; foreword by Dean Kamen ; afterword by Woodie Flowers.
        p. cm.
    ISBN-13: 978-1-59253-411-1
    ISBN-10: 1-59253-411-2
    1. Robots--Design and construction. 2. Robotics. 3. FIRST Robotics Competition (2007) I. Slezycki, Stephanie. II. Title.
    TJ211.W415 2008
    629.8'92--dc22

                                                2007036611
                                                CIP

ISBN-13: 978-1-59253-411-1
ISBN-10: 1-59253-411-2
10 9 8 7 6 5 4 3 2 1

Design: Kathie Alexander

Printed in China

FIRST Robots:
Rack 'n' Roll

BEHIND THE DESIGN

30 Profiles of **Award-Winning** Robot Design

Vince Wilczynski
Stephanie Slezycki

Foreword by Dean Kamen
   Founder, FIRST

Afterword by Woodie Flowers
   Cofounder, FIRST Robotics
   Competition

BEVERLY MASSACHUSETTS

ROCKPORT PUBLISHERS

Dean Kamen

Inventing and creating are very distinct processes and each is extremely rewarding.

Inventing involves looking at an unsolved problem and using one's imagination to think of possible solutions. Investigations ensue and some solutions prove to be promising, while others not. In due time, an optimal solution is identified.

Creating is the process that brings objects into being. Materials are modified and combined to transform a collection of components into a working solution. The only practical inventions are those that can be created.

Inventing and creating are important aspects of FIRST. They are the hooks that attract individuals to a much wider and important purpose. Teams are presented with a design challenge, and they rely on their creative abilities to invent possible solutions. A short period of testing and evaluation follows during which time the best solutions rise to the top and become the team's robot. FIRST is a fast-paced project, but its impact is long lasting.

The real magic in FIRST is not what happens with the robots but, rather, is the transformation of FIRST team members. In fact, it is the process of invention and creation—and not the product itself—that is most important in FIRST.

On FIRST teams, students work side by side with engineering mentors to design and build robots. This interaction between the mentors and students is a powerful catalyst for promoting an interest in technology among the students. By sharing their passion for engineering, the mentors demonstrate that a career in engineering is immensely satisfying. The students are immersed in real-world problem solving and they expand their horizons for future possibilities.

Because FIRST teams are similar to small businesses, participants are exposed to a wide range of experiences, including fund-raising, public relations, computer animation, marketing, and journalism. Mentors join FIRST teams to document and expand the teams' influence on societal views about science and engineering. Students are exposed to a variety of career fields and experience the powerful impact that can be achieved when diverse groups of committed individuals work toward a common goal.

Inspiration is the essence of FIRST. The robots and competitions are exciting and fun, but the mentors use these tools for a much greater purpose—to inspire students. FIRST mentors are special people not only because of their demanding careers solving the world's problems and improving our standard of living, but also because of their willingness to give of themselves for the benefit of others.

In 2007, there were 1,300 FIRST Robotics Competition teams from seven countries. More than 32,500 high school students participated in FIRST, where they competed at thirty-seven differ-

▲ Dean Kamen is president of DEKA Research and Development Corporation, a dynamic company focused on the development of revolutionary new technologies that span a diverse set of applications. As an inventor, physicist, and entrepreneur, Kamen has dedicated his life to developing technologies that help people lead better lives. One of Kamen's proudest accomplishments is founding FIRST.

ent events in Brazil, Canada, Israel, and the United States. The FIRST Championship, held in Atlanta, Georgia, brought together more than 300 of the top teams for the premier competition.

FIRST competitions include a number of awards that promote creativity, design excellence, controllability, and quality. At each competition venue, a panel of judges examines each robot against detailed award criteria. The judge panels are well suited for scrutinizing each design since the judges are the engineering, technology, science, business, and government leaders within the community. Even through the judging process, students are exposed to role models.

While this book documents some of the best robot designs from the 2007 FIRST Robotics Competition, it is essential to remember that the robots are simply artifacts of a process that was created to develop people. The simplicity and effectiveness of the reported creations are a compliment to the talent of the inventors responsible for these designs.

Nearly 200 FIRST teams were recognized in 2007 for their engineering expertise at one of our competitions. From that set of award-winning teams, thirty teams were selected as exemplars and are profiled in this book. Needless to say, these teams are some of our finest. By celebrating the designs, we are celebrating and complimenting the designers.

Winning on the competition field is not the ultimate goal in FIRST—winning at life is. FIRST is an invention that partners mentors and students to create new futures. I hope this book gives you some insight into the possibilities of FIRST.

If you are not yet involved in FIRST, I invite you to do so. You will find the experience inspiring.

◄ By challenging teams to hang inflated tubes on a center goal, the 2007 FIRST Robotics Competition, called "Rack 'n' Roll" required teams to design complex robots having quick speed, humanlike dexterity, and a long reach.

▶ A full rack of tubes was the ultimate goal for each alliance, with the more tubes in a row, the better. A complete ring of eight alliance tubes was worth 256 points—a score that was hard to beat.

2007 **FIRST** Robotics Competition: **Rack ·n· Roll**

One of the most amazing aspects of the FIRST Robotics Competition is the compressed forty-five-day design cycle. Entire robots and all of their integrated mechanisms, components, electronics, and computer control programs are conceived and brought into being during a six and a half week period.

Initial concepts are sketched as soon as the new game for the year is unveiled in January. The design pace accelerates as ideas are tested, prototypes fashioned, and intricate mechanisms fabricated. The design process comes to a screeching halt in the middle of February when each team packs their creation into a crate and ships the robot to a competition venue. The team is reunited with their robot the first morning of the three-day competition to make final improvements and bolster their chance for success.

A new game is created each year and kept as a surprise until its unveiling at the start of the season. An international broadcast announces the year's challenge and the rules governing competition play.

SCORING IN RACK 'n' ROLL

The 2007 game consisted of two distinct methods for scoring points: picking up and placing tubes on a center goal, and climbing on top of a partnering robot. These very different scoring possibilities required teams to analyze the game and design a robot that either favors a single scoring strategy or attempts to accomplish both scoring tasks.

An ominous looking eight-legged structure dominated the middle of the playing field. This rack had three levels of scoring stations, with eight legs on each level. Points were awarded for each tube hung on one of the legs on the rack. The legs were pinned together at the rack's center and suspended from chains attached to the top of the rack. A small circular disk, called the foot, was attached to the end of each leg to prevent the tubes from falling off once they scored.

Because loose connections suspended each leg, the scoring target would swing horizontally when impacted by a robot. The movement of an individual leg was translated to the other legs since pinned connections linked the entire structure together. Though each leg was a rigid length of pipe, the composite structure was compliant and swayed easily. Robot interactions with the goal caused the entire suspended structure to swing in unpredictable patterns, thereby providing the basis for the competition name Rack 'n' Roll.

EXPONENTIAL POWER OF THE TUBES

During competition play, three teams were randomly partnered as an alliance for each match. The alliance worked together to gather the

▼ With similarities to a spider, each of the eight legs radiating from the center of the rack was a target to hang tubes on. A circular plate at the end of each leg, humorously named the spider's foot, ensured that tubes scored on the legs stayed in place.

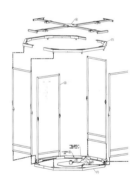

32-inch (81.3-cm) -diameter tubes and place them on a leg of the rack. Rack scoring was a function of the number of consecutive alliance tubes in a row. A single tube on the rack was worth two points and each successive tube multiplied the row value by a factor of two. For example, the scoring algorithm awarded four points when two alliance tubes were next to each other, sixteen points if four alliance tubes formed a row, and 256 points if an alliance had a tube on each of the eight spider legs in one level. The progressive value of placing tubes in continuous rows dictated the alliance strategy to maximize scoring.

Each leg had only enough room to hold two tubes and, with one exception, only the first tube on a leg counted when scoring a match. Three types of tubes were named based on their scoring capabilities. Ringers were solid-colored red or blue tubes that could be placed only on, but not removed from, the scoring legs. Spoilers were black tubes that nullified the value of a tube behind it on the rack. Unlike the other tubes, spoilers could be removed from the rack by the opposing alliance. Blue and red FIRST-labeled tubes were called keepers. Once a keeper was placed on the rack, no other tubes were allowed on that leg, thereby preserving that position for the respective alliance.

Keepers could only be scored during the first fifteen seconds of the match in the period called autonomous mode. During this period, the robots operated according to preprogrammed instructions. Sensors, such as a camera to detect the green lights placed on top of the rack, detected the robot's environment and motions. These signals were interpreted by the onboard microprocessor, with the program then executing robot functions based on the sensor measurements. The autonomous mode was followed by two minutes of driver-controlled play.

CLIMBING TO RACK UP POINTS

Climbing on top of an alliance partner was the second form of scoring. Teams earned thirty points if they were on an alliance partner and at least 12 inches (30.5 cm) off the floor and fifteen points if the robot was 4 inches (10.2 cm) off the ground. These scores were awarded only when the robots were parked in the far ends of the field in areas called the home zone for each alliance. Since the

scores were determined at the end of the match, the final seconds of competition included high levels of excitement as robots abandoned attempts to score tubes and scrambled to their home zone to climb on an alliance partner.

TUBE-BOTS, RAMP-BOTS, AND HYBRIDS

The unique scoring options prompted three types of robots. Tube scoring robots focused on the ability to grab tubes off the floor and place them on the rack. The most successful "tube-bots" included arms or elevators that enabled the robot to score on all three levels. "Ramp-bots" were designed to serve as platforms for other robots to climb on at the end of the match. Since size constraints restricted each robot to a 28 × 38 inch (71.1 × 96.5 cm) starting footprint, ramp-bots began each match with their climbing platforms folded up on the robot. Near the end of the match, the stored platforms would be released and expand to create large ramps and robot parking lots.

Robots that both scored tubes and supported alliance robots were called hybrids. These robots were often the most complicated machines as teams optimized multiple components to achieve the dual functionality and still meet the competition's size and weight restrictions for each robot.

The multiple methods for scoring resulted in matches where strategy became an important aspect of the game. Players and coaches kept a close watch on the rack to determine the optimal placement for each tube. Scores advanced quickly as teams extended rows of scored tubes, but the momentum immediately shifted if a spoiler made its way onto the rack.

In the final seconds of the match, alliances evaluated their positions to decide the relative value of adding more tubes to the rack, defending their scored tubes from the spoiler, or racing back to their home zone to climb on an alliance partner. The game required not only a well-designed robot but also a skilled set of coaches and players that could constantly survey the field, estimate the current score, make quick judgments, and ultimately win the game of Rack 'n' Roll.

▲ Scoring a tube on the rack was a complex task, requiring robot drivers to grab a tube off the floor, position the robot near the rack, lift the tube to scoring position, navigate the tube past the foot, and finally release the tube onto the leg.

▲ The FIRST label identified keepers as special tubes that could only be scored during the fifteen-second autonomous period at the beginning of each match. Onboard sensors provided input to the robot controller to determine the robot's position relative to the scoring rack.

▲ Winglike appendages sprouted from ramp-bots to create platforms for alliance partners to climb on at the end of each match. Because robots that climbed the highest ramps earned thirty points, the results were well worth the effort.

◀ The Kit of Parts includes mechanical, electrical, and control components that can be used to build a FIRST robot. A 12-volt battery serves as the main power source for the robot.

Building a **FIRST** Robot: **Kit of Parts**

THE FIRST ROBOT RULES

FIRST robots are constructed according to a set of rules that details the components that can be used, explain safe operating conditions, and specify the periods of time when the robot can be constructed. The robots must be assembled from the components included in the 2007 FIRST Kit of Parts and other materials specified in the rules.

The robots must be constructed within a six-week window that follows the announcement of the game in January. Before this period, teams are free to design robotic mechanisms, with the understanding that they are doing so without knowing the specific game they will need to play. Also, while designs developed before the game announcement can be used, the actual components must be constructed and assembled during the specified fabrication window. Similar rules apply to control software such that the software installed on the robot must be written during the six-week fabrication window.

FIRST robots must be fabricated or assembled from items that are constructed from materials provided in the kit of parts or from a list of additional materials listed in the rules. The total cost of all non-kit components must not exceed $3,500 USD and no single component can have a value of over $400. Components can also be recycled from previous years' robots, but their original cost must be accounted for.

To promote safety, teams are encouraged to use bumpers on their robots. The bumpers must meet the specified construction standards and are required to be mounted in a specific vertical position on the robots. The rules favor robot-to-robot

The kit includes twelve motors to power the individual systems on a FIRST robot. Most motors require a transmission be added to the motor output shaft to increase the torque to a useful level. Some motors in the kit include a transmission, while others do not.

interaction in the bumper zone and penalize teams for collisions outside the specified zones.

In addition to components and work periods, the energy used by FIRST robots is also regulated. The electrical energy is limited to that supplied by the 12 V power battery and the 7.2 V control system battery. The compressor provided in the kit must be used to compress air which is stored in a limited number of storage cylinders. The final form of allowable stored energy is the deformation of parts, such as springs.

THE KIT OF PARTS

The 2007 FIRST Kit of Parts provides the backbone for constructing a FIRST robot. The kit includes enough material to construct a fully functioning robot base, including the power source, electronics, and control system. In addition to these parts, the kit includes a wide assortment of mechanical and electrical components, as well as software. Many companies donate material to be included in the kit as a means to support FIRST while getting their product into the hands of the world's top engineers.

Certain aspects of the robot must be constructed using only the material in the kit. Specifically, the motors provided in the kit are the only motors that can be used on the robot, and the control system to operate the robot must only use components provided in the kit. If a team chooses to use pneumatic power, that system must be assembled from material provided in the kit.

MOTORS AND CONTROL

Twelve 12 V, direct-current motors are provided in the kit. Up to three additional motors, identical to the two most powerful motors provided in the kit, can be included on a robot. Some motors are shipped with a transmission to reduce the output speed and increase the output torque.

The motor transmissions can be altered, but the motors must be used in their original state. This rule preserves the integrity of the motors and ensures that they are safely used in their designed configurations. In addition to these motors, an unlimited number of low-power servo motors may be used on the robot. A common use of these motors is for tracking systems and to release or activate latching mechanisms.

The control system is designed and manufactured by Innovation First exclusively for FIRST Robotics teams. The major components of the control system include an operator interface, a robot controller, a transmitting radio, and a receiver.

The robot controller has a variety of functions. Its main purpose is to translate the commands sent from the operator interface into commands that initiate robot actions. The controller also has the ability to collect and interpret data from sensors mounted on the robot. Since the controller is programmable, specific functions can be executed based on the signals measured by the sensors.

A common use of the controller in the 2007 competition was to interpret output from a robot-mounted camera that detected the position of the target light. The controller interpreted the data to actuate the drive motors and steer the robot toward the target light. Another common use of the controller included closed-loop control algorithms to

▶ A planetary gear transmission was included in the kit to reduce the output speed of the most powerful motor. A long output shaft offered ample room to mount wheels and sprockets, as long as it was properly supported with an additional shaft bearing.

▼ Four wheels were provided in the kit. The wheels were specifically designed for FIRST robots with features such as hubs with predrilled holes for attaching sprockets and soft treads with a high coefficient of friction.

▶ All power leads on the robot were connected to a terminal block and individually protected by fuses. The terminal block was expandable by adding additional sections to accommodate the leads for each motor.

monitor the exact locations of robot arms to score tubes on the goal.

When a driver activated a joystick or push button, that signal was received and coded by the operator interface. The signal was transmitted as a radio signal and received by robot controller. The driver's command was interpreted by the robot controller and converted into the control function needed to create the commanded function. This process was handled seamlessly and flawlessly by the Innovation First control system.

PNEUMATIC COMPONENTS

A complete pneumatic system including an air compressor, storage tanks, connecting tube, pistons, pressure gauges, regulators, solenoid valves, and flow-control fittings was included in the kit. Additional pistons could be ordered to achieve the range of motion and force required in a specific design.

The pneumatic system also included a vacuum generator and vacuum cups. The vacuum generator operated using a venturi effect where pressure drops as fluid flows through a restriction. The vacuum generator produced sufficient vacuum to hold a tube and offered the ability to immediately release the tube by ending the vacuum.

Solenoid valves were provided as part of the pneumatic system to activate air cylinders. The solenoid valves were triggered by the control system and directed high-pressure air to open and close pistons. Many teams used pneumatic pressure to operate manipulators to control the tubes. These systems operated quickly and were highly reliable—two attributes of pneumatic-powered mechanisms.

ADDITIONAL COMPONENTS OF THE KIT OF PARTS

The kit also included structural members to construct a robot frame, as well as the components to build a drive system for the robot. Heavy-gauge aluminum rails could be assembled in a variety of configurations to serve as a base for the robot. Drive motor transmissions, sprockets, wheels, and chain were included in the kit to assemble a basic means to propel the robot.

Other components of the kit included bearings, wire, sensors, fuses, and other hardware to build the robot. In addition to the hardware, the kit also included software. National Instruments LabVIEW data-acquisition software was provided to enable teams to evaluate sensors and tune robot cameras for optimum performance. Microchip Corporation's MPLAB CBOT computer-programming compiler was included in the kit for writing, debugging, and compiling the C-code that controls the robots. Intelitek's easyC PRO software, an icon-based programming environment, was also included in the kit.

Though the kit could be used to construct a basic robot, another important purpose of the components is to prompt problem solving. The raw materials in the kit give teams objects to handle and ponder as they imagine mechanisms that achieve the game function. The kit parts provide a window into the possibilities, with the teams building from this starting point to create their own unique mechanisms and robots.

▶ The robot control system included a controller, receiving radio, voltage controllers, and voltage relays. Sensors were also an important component of the control system on many robots.

The scoring tubes were perhaps the most important part in each kit. By having the tubes available at the start of the design process, accurate and reliable methods for manipulating these objects could be created.

The camera was the most sophisticated sensor included in the Kit of Parts. Once calibrated, the camera could track the target light that marked the location of the rack's spider legs.

Delphi Driving Tomorrow's Technology Award

The Delphi Driving Tomorrow's Technology Award celebrates an advantageous machine feature that demonstrates any aspect of engineering elegance, including design, wiring methods, material selection, programming techniques, and unique machine attributes. The machine feature must be integral to the robot's function and make the team more robust in competition.

Delphi is a global supplier of mobile electronics and transportation systems, including powertrain, safety, steering, thermal, controls and security systems, electrical/electronic architecture, and in-car entertainment technologies. As a founding sponsor of FIRST, Delphi has been instrumental in its growth and success. Generously supporting regional events and many teams, Delphi has given young people the opportunity to learn from and play with the pros in the engineering world. Its long-term support of FIRST has also given its engineers the exciting, reenergizing opportunity to help students learn engineering skills, as well as form lasting connections with them from high school through college, internships, and into the workforce. The former chairman and CEO of Delphi has served as a FIRST board member.

An Integrated Approach to Drive Systems

▼ Team 118, the Robonauts, presented a creative new approach to the FIRST robotic drive system, demonstrating how they are driving tomorrow's technology.

The 2007 Rack 'n' Roll game not only presented a design challenge, but provided Team 118, The Robonauts, with the inspiration to reexamine the concept of a drive system. In their eleventh year of FIRST involvement, this Houston, Texas, team enhanced previous robot design features while rethinking the traditionally contiguous motor/transmission combinations. Team 118 achieved both the power to fend off opponents and the speed and agility needed to navigate the playing field by undertaking a unique approach to drive system integration.

EVOLUTION OF THE V6

The design for the V6 drive system came about through an evolution of ideas, stemming from the brainstorming sessions at the start of the season. The team began designing their machine to have a low center of gravity. Initial sketches positioned the heaviest components of the robot on the base of the frame, including three separate drive motors with one transmission each. With the battery and other components also taking up space, there was no room left for a desired fourth motor/transmission combination. The combined units' weight was also nearing a dangerous level to comply with the maximum allowed specification.

During deliberations as to whether a shifting mechanism was necessary, or how it could be better incorporated, a suggestion was presented to create a system of one large bevel gear to drive a single DeWalt kit transmission, with the drive motors mounted around and driving that single gear. This proposal soon evolved, and the bevel gear was replaced with a system of spur gears, mounted in close proximity and driving a central input gear that would directly connect to a single transmission.

As the design began to form in more detail, the team decided a CAD model would be essential in determining the feasibility of such a system. Initially, the four CIM motors included in the Kit of Parts would be the main driving motors, but two Fisher-Price motors were added to further increase the power output of the system. The computer model revealed that the prospective motor/transmission combination was indeed possible to build, and work immediately began on a prototype.

An initial motor system was constructed and installed on a practice robot, with the intention of driving the machine to failure. If the drive system did fail, the team would be able to conclude early on in the build phase if a new approach would be needed. However, if the system proved to be functional and practical, the tests would reveal any glitches in the system that would need to be worked out.

The trial robot was driven hard, and many potentially damaging scenarios were played out. The machine was driven at full speed, driven into walls, and made to push heavy weights across the floor. The motors and transmission passed every test. Team 118 had found the key component that would make their robot stand out in competition.

▶ With the drive motors and main battery on opposite sides, the robot maintained a low center of gravity.

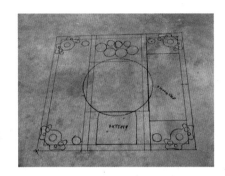

◀ Three-dimensional modeling software enables the team to visualize the integration of robot components prior to manufacturing. This helps prevent possible element interference and can aid in structure compliance early in the design process.

IMPROVING THE CONCEPT

As the system coupled six of the highest power output motors in the kit with a single shifting transmission, it was given the nickname of V6. By connecting the drive motors and the transmission into one unit, they no longer had to be spread out along the robot's drive wheels. This freed up space on the base for other components, allowing more of the total robot weight to be focused on the lowest part of the machine.

Work began on revisions to the prototype drive. The drive system was fitted with a pneumatic braking device that, when activated, would assist in preventing the robot from being pushed by an opponent while scoring. To avoid the activation of the brake system while the robot drove at high speeds—which could destroy the transmission—the controlling program was written to include a feature that would disallow the brakes to be fired within one second of the application of voltage to the motors. A slip clutch was also installed to allow the robot to slip free if the brakes were fired while the robot was moving but also provide enough force to stop the robot from being pushed.

Other improvements included the replacement of the gear axle bushings with high-speed bearings after they showed noticeable wear. A torque plate was added to divert forces away from the transmission housing and into the robot's frame. Gear spacing was optimized, and unnecessary material was removed to increase the overall efficiency of the system.

A final CAD model was made to resolve any further integration issues and incorporate these changes, and three complete V6 models were manufactured from the CAD model. One was used on the practice robot, and the other two were taken to the competition, with one reserved as a spare.

▶ The prototype of the V6 endured harsh testing, including the 1600 watts of power applied from the motors to the carpeted floor. The system proved successful and was an important part in the Robonauts' 2007 machine.

▶ The improved unit includes four CIM motors and two Fisher Price motors. The spur gears can be seen for each of the motors, with the four on the lower circumference matching the CIMs, the smallest at the top matching the Fisher-Price motors, and a central gear to connect the six motors to the DeWalt transmission. The ratio between the CIM motors and the input shaft is 1:1, and the ratio between the Fisher-Price motors and input shaft is 3:1.

◀ The final model of the V6 consists of four CIM motors, two Fisher-Price motors, one DeWalt transmission, a pneumatic brake system, and a servo shifter. This 3-speed unit provides a top speed of over 16 meters (52 ½') per second, and a pushing force exceeding 300 lb.

▶ A view from under the robot shows the mounted V6 drive system. A mirror image of the V6, manufactured as a spare, is shown across from the installed unit. These systems are modular, allowing rapid access for repair or replacement.

SIMPLIFICATION BY INTEGRATION

The V6 drive system was mounted on the robot base and connected to a crab drive used by Team 118 on previous robots. The crab drive allowed the wheels to rotate along a vertical axis, permitting the robot to move in any direction without changing its physical orientation.

The completed robot, named Redline, incorporated fully connected steering and drive systems. The four wheel-boxes on the robot were connected and linked to a single sensor assembly, guaranteeing simultaneous rotation. The design enabled the robot to rotate 360 degrees in less than one second, so the motors needed to spin in only one direction. The elimination of change in motor direction from forward to reverse aided in reducing transmission wear. This also provided a smooth transition of direction; instead of the motors changing their direction of rotation, when the driver pulled the joystick from forward to reverse, the motors continuously turned as the wheel-boxes spun 180 degrees.

A system of connected drive chains made the synchronization of the drive wheels possible. The four wheels spun at the same speed, eliminating the occurrence of drifting or unbalance, and if a single motor or chain were to break during competition, the remaining ones would continue to function. The connected drive design meant the V6 motor arrangement could be positioned along any point of the drive system, allowing more flexibility in the special layout of the robot base.

The V6 system fulfilled both the power and speed requirements needed by Team 118. The resulting machine could travel at a top speed of more than 16 meter (52.5 feet) per second, while also able to apply a pushing force of more than 300 lb.

The integration of the six drive motors with a single transmission provided a versatile, unitized powerhouse that greatly reduced the limited physical area and weight that a system of separate motors and transmissions would have consumed. An innovative look at the integration of drive system technology equipped the Robonauts with a smooth-driving machine that mastered the playing field.

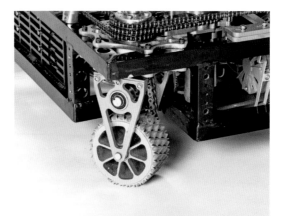

◀ This complex crab wheel is one of the four used to translate the robot in different directions without a realignment of the base. As with the rest of the robot components, these were first fully modeled in Pro/E CAD software before being manufactured.

◀ Power is transmitted from the blue drive chains down the vertical shaft to the crab drive wheel by means of bevel gears. The green chains rotate the wheel module continuously to change the robot's direction.

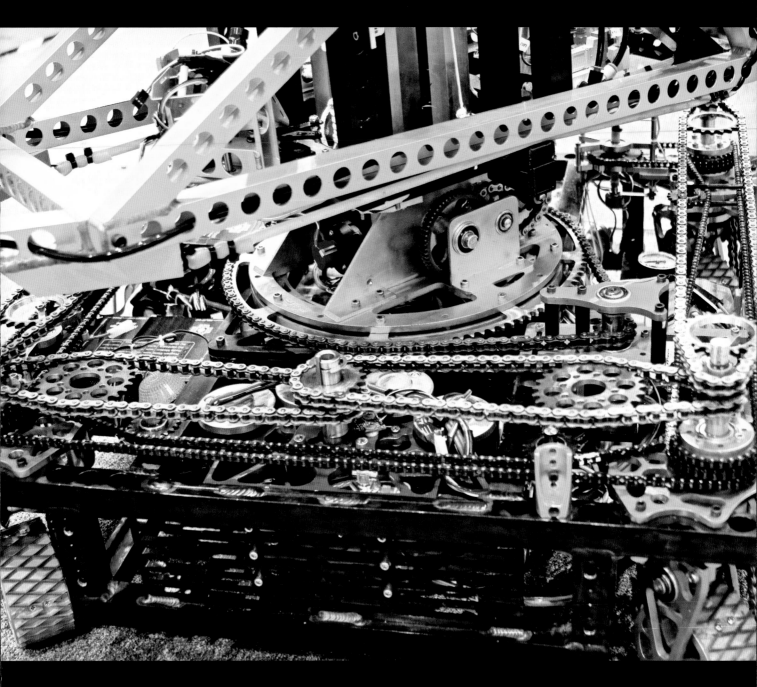

▲ This complex robot is driven by forty-six sprockets and ninteen gears. More than 35 feet (10.6 meters) of drive chain is broken into seventeen individual loops, enabling redundancy and synchronizing the wheel boxes.

◀ The V6 motors and crab drive system, when coupled with an effective tube manipulator, create a highly competitive machine, able to easily navigate the playing field to score points.

▲ Team 118 successfully integrates their crab drive with the innovative V6 design. Together, these components help carry the team through their successful competition season.

Incorporating Extensive Mobility for Success

Team 173, Robotics and Gadget Engineering (aka/RAGE) from Hartford, Connecticut, has many years of experience with FIRST competitions. For the 2007 season, they took on the Rack 'n' Roll game with very specific requirements in mind. They set basic goals, and modified their robot design to fit within these parameters. The result was a constantly evolving and improving device for maneuvering game pieces from the playing field floor to the center rack. This device incorporated intricate methods of handling the pieces to maintain control of them throughout each match.

Team RAGE opted to build a Class I robot, which was allowed a footprint of 28" × 38" (71.1 × 96.5 cm), a maximum starting height of 48" (121.9 cm), and maximum weight of 120 lb. Although this is the smallest physical class, it allows for the most machine weight. The height of the tallest spider on the rack was 92" (233.7 cm), almost twice the height of the robot's starting configuration. It was at this point that the team determined the main focus of the robot would be on extending itself to score on the highest rack. With this capability, along with a reliable method of picking up and moving game pieces, the robot would be able to score at any of the three rack levels.

To reach and score on all three rack heights, the conceptual robot would need a gripping mechanism attached to an arm that, at the beginning of each match, would be folded to fit within the Class I size restraints. A pivoting arm with a tilting gripper head began to form as prototyping commenced.

▲ Team RAGE designs a machine that is capable of precisely handling the game pieces, allowing a controlled approach to manipulating them from the playing field floor to the rack for scoring.

DESIGN CONSIDERATIONS FOR TELESCOPING PARTS

The arm was constructed out of three different-sized tubes of rectangular extruded aluminum, which fit inside each other when contracted and extended out when expanded. Extruded aluminum was chosen for its high strength-to-weight ratio. A chain-driven, 17" (43.2 cm)-diameter turntable was mounted to the robot base, and to this attached the largest of the telescoping tubes at a pivot. This turntable and pivot enabled an increase in the range of motion and angle of elevation of the gripping device.

The bending stress and maximum load conditions were analyzed to determine the size of the tubing needed to withstand the pressures of play. The material had to be hollow to account for the other telescoping tubes, and had to mesh well with the stock bearings and collar clamps that would connect it to the robot. Different types of extrusions were considered to minimize the overall cross section of the arm, while still providing enough free space to run the Igus Energy Chain, which housed the wires and pneumatic lines that would control the gripping mechanism.

The slide bearing loads as a function of tube extension were analyzed to find the necessary tube overlap, and the loads on the selected drive motors for the turntable, tube extension, and angle tilt were analyzed to ensure viability. Bearings fabricated from ultra high molecular weight (UHMW) polyethylene were used to provide the telescoping parts smooth motion. End stops of the same material were included on the end of the bearings to prevent the overextension of the tubes' linear movement. The team found that hard stops imparted a large impulse into the system when impacted, potentially causing enough physical shock to damage the mechanism. An integral spring stop system was designed to prevent such damage.

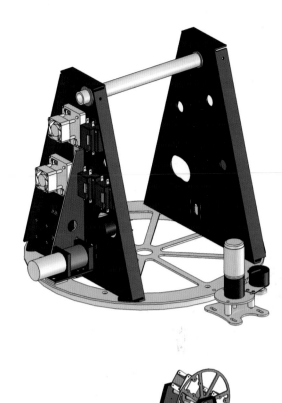

▶ Stanchions are mounted on the rotating turntable to support a pivot shaft with a large sprocket. The telescoping arm is clamped to the sprocket, which controls the tilt angle.

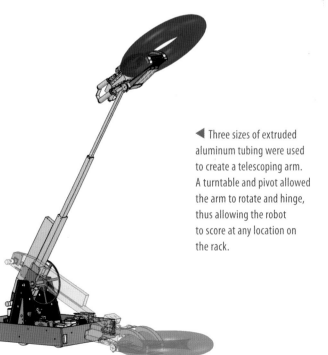

◀ Three sizes of extruded aluminum tubing were used to create a telescoping arm. A turntable and pivot allowed the arm to rotate and hinge, thus allowing the robot to score at any location on the rack.

Example Drawing to JP Fabrication

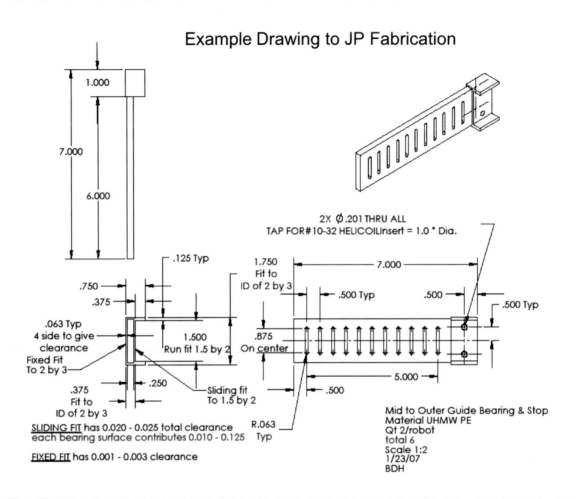

1.000

7.000

6.000

.125 Typ

.750

.375

.063 Typ
4 side to give
clearance
Fixed Fit
To 2 by 3

1.500
Run fit 1.5 by 2

.375
Fit to
ID of 2 by 3

.250

Sliding fit
To 1.5 by 2

SLIDING FIT has 0.020 - 0.025 total clearance
each bearing surface contributes 0.010 - 0.125

FIXED FIT has 0.001 - 0.003 clearance

R.063
Typ

2X Ø .201 THRU ALL
TAP FOR #10-32 HELICOIL Insert = 1.0 ° Dia.

1.750
Fit to
ID of 2 by 3

7.000

.500 Typ

.500

.500 Typ

.875
On center

5.000

.500

Mid to Outer Guide Bearing & Stop
Material UHMW PE
Qt 2/robot
total 6
Scale 1:2
1/23/07
BDH

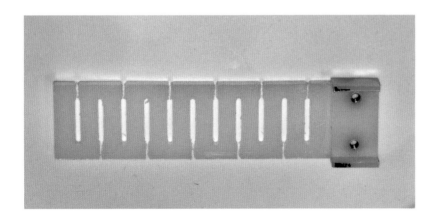

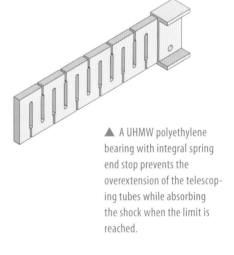

▲ A UHMW polyethylene
bearing with integral spring
end stop prevents the
overextension of the telescoping tubes while absorbing
the shock when the limit is
reached.

SPACE RESTRICTIONS AND LINEAR TO ANGULAR MOTION

At the furthest end of the telescoping arm was a game-piece gripping device. Initial designs included suction cups to attach to the game piece surface, designs to grip the piece from the inner diameter, and wheels to grasp the pieces from their edges. The inside grabber interfered with game piece placement on the rack, and the wheels were difficult to fit within the starting size restriction. Through testing, these ideas eventually evolved into a dual-jawed gripping mechanism that could tilt to three positions to allow the lifting of game pieces from the floor and placement on the three rack levels.

To fit inside the starting envelope, the jaws would need to be in a fully open position, about 135 degrees from the closed position. Pneumatic linear actuators were installed to control the jaw movement. To convert the linear motion of the actuators into a rotation, a chain was attached to the actuator, which wrapped around a chain wheel and rotated each jaw at the hinge. To return the chain to its original position, a latex tube was stretched to act as a tension spring.

The two jaws of the gripping mechanism were designed to act independent of each other. Initially, the required force to hold the game pieces was calculated to determine the specific design. A force gauge was used to measure the minimum required force for the two jaws to maintain a constant hold on a single game piece in a horizontal configuration.

▶ The starting position of the robot requires the jaws to be rotated 135 degrees from the closed position. A chain attached to a chain wheel converts linear motion of a pneumatic linear actuator into the rotation needed to close the jaws.

◄ A prototype of two jaws is fitted with a force gauge to determine the amount of force needed to prevent a game piece from slipping as different friction material is applied to the inside surface of the jaws. The required force to firmly hold the piece is approximately 4 lb.

UNBALANCING THE FORCES

Team 173 soon learned that the application of equal force between the top and bottom jaws led to an unpredictable balance of the game piece between them. The position of the piece was important in providing accurate placement on the rack. To regulate position, the forces applied to each jaw were modified. The lower jaw applied a larger torque, so the weaker upper jaw would flex as the bottom jaw approached a fully closed position. This difference in torque was applied through the use of different diameter sprockets and pistons and resulted in a lower jaw having 50 percent more force than the upper jaw. To reduce the occurrence of slippage between the chain and sprocket, the sprocket was replaced with a chain wheel, with the chain anchored to a lug on the wheel. Pistons with large diameters and short strokes enabled the use of smaller diameter chain wheels, helping the system fit within the starting size requirements.

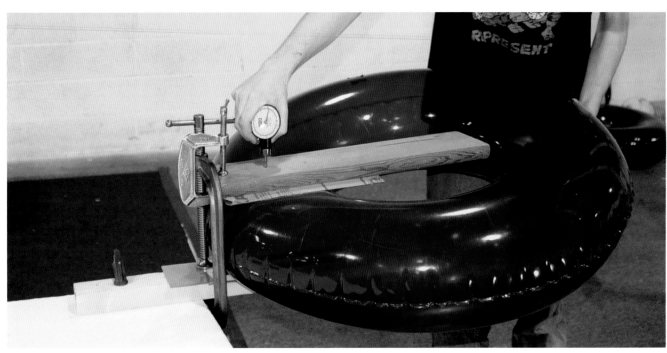

▶ Different-sized chain wheels and pistons control the upper and lower jaws of the gripper head. One end of each chain is connected to a piston rod and the other to a stretched latex tube to keep tension on the system.

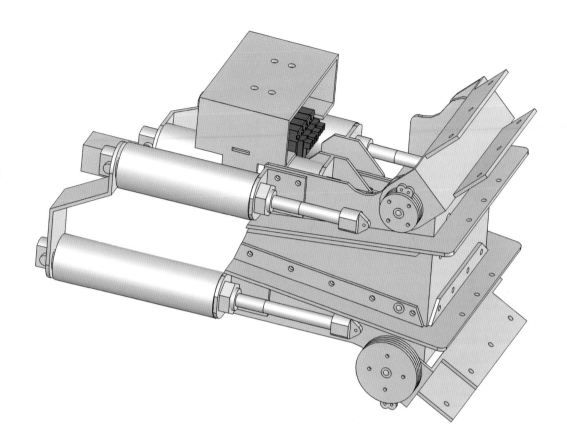

THE FRICTION FACTOR

▼ A curved upper jaw and wedge-shaped lower jaw closed to pick up and maneuver game pieces. The upper surface of the lower jaw has a flexible, slotted spring plate that is deflected when a piece is grabbed, allowing contact with high-friction material.

The manner in which the game pieces were held was not only dependent on force, but also on the friction applied by the jaw material. A curved upper jaw featured high-friction material that evolved throughout the season. Several materials were used, including a room temperature vulcanizing (RTV) silicone rubber that did not adhere well, a combination of two adhesives—which took a long time to cure and rapidly accumulated dust that decreased the friction factor—and precast RTV hemispheres. None of these materials provided the required grip needed to successfully maneuver game pieces. The final material was borrowed from Team 190, who showed gracious professionalism in providing Team 173 with the high-friction material used on their machine.

The wedge-shaped lower jaw posed a larger problem regarding the friction requirements. It needed to be smooth enough to slide under game pieces on the floor without catching the carpet and rotating the gripper head, but also have equal friction to the upper jaw for carrying the pieces. The resolution came in the form of a flexible slotted spring plate that covered the lower jaw. The plate could easily slide along the carpeted playing field floor, and when the jaws closed, the spring plate deflected to expose the traction material through the slots. Thus, the plate provided the lower jaw with both smooth and rough surfaces.

Through constant improvement of the gripper design, Team 173 produced an ever-evolving machine. They continuously researched different design methods and materials to ensure they found optimum robot features. Their drive toward learning is evident in their constant quest for enhancement.

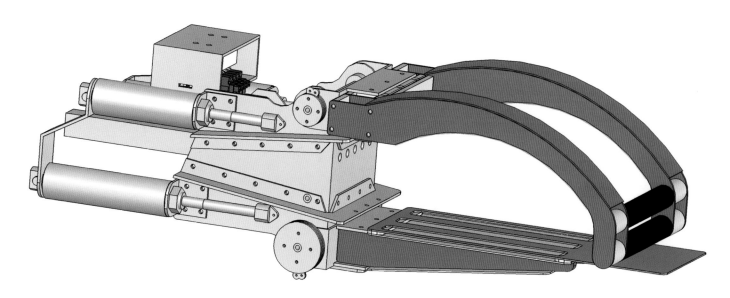

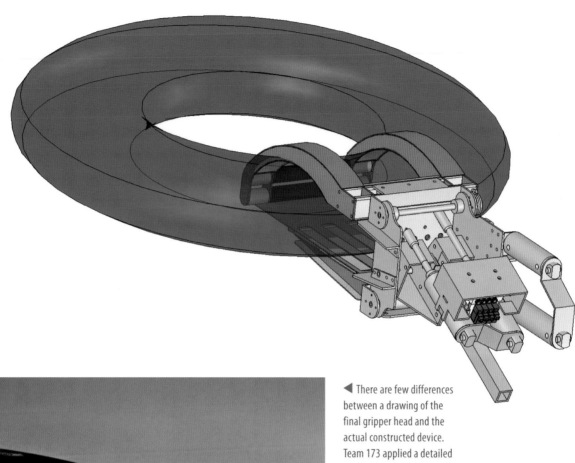

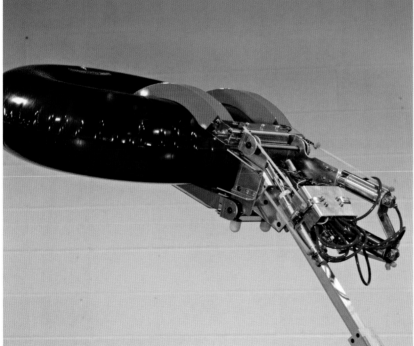

◀ There are few differences between a drawing of the final gripper head and the actual constructed device. Team 173 applied a detailed approach to the design of their award-winning feature.

Designing an Inertial Platform

Sensor-guided robots have a great advantage when it comes to scoring, especially in the autonomous rounds. The CMUCam2 vision sensor can find and lock onto the illuminated targets at the top of the rack, potentially scoring points when robots unequipped with such sensors sit idle. When programmed correctly, additional sensors can greatly enhance a team's strategy and performance. Team 375 of Staten Island, New York, applied their technological know-how to create an impressive and advantageous array of sensors.

COMPILING A SENSOR ARRAY

An inertial platform is essentially an electronic system that provides guidance to a vehicle through use of continuous monitoring of position, velocity, and acceleration. Guidance algorithms and other control mechanisms convert measurements into useful data to navigate the machine.

The inertial platform on Team 375's robot consisted of an array of sensing devices used to support the inertial guidance system. Initially, the system consisted of only a single vision sensor, the CMUCam2 provided in the Kit of Parts. The team found that one sensor did not provide high enough accuracy for the desired robot performance. Other design alternatives considered were motion sensors and a sensor-guided dead reckoning system using ultrasonic range finders.

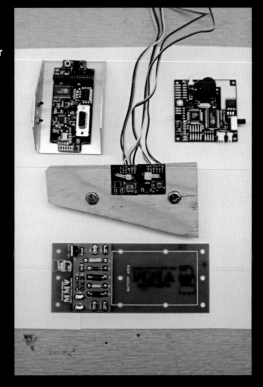

◀ The sensors that comprise the inertial platform include two CMUCam2 vision sensors, an electronic yaw rate gyro, and a dual axis accelerometer. Also shown is the backup battery charging circuit that powers the cameras and associated pan/tilt assemblies.

To obtain accurate field navigation, additional sensors were incorporated into the design, then integrated and programmed with a control algorithm. Along with camera vision sensors, a yaw rate gyro and dual axis accelerometer were incorporated to provide position feedback, giving the guidance system an accuracy rating of 20 seconds of arc from a distance of 30 feet (9.1 m).

Testing and prototyping of the sensor array was performed to integrate the feedback from each sensor into a single set of control loops. Fine-tuning of these control loops completed the synthesis to create a fluent sensor system.

▲ Team 375's robot has positioned itself in front of the rack using the inertial platform. A game piece is about to be scored in autonomous mode.

▶ The two independently moving camera eyes give the appearance of a chameleon. The use of two cameras equips the robot with accurate navigational ability.

CHAMELEON VISION

Two CMUCam2 cameras were mounted on a crossbar near the top of the robot, separated by approximately 10'' (25.4 cm). These cameras functioned independently, and could pan and tilt in different directions at the same time. This provided the ability to run the cameras using a stereoscoping algorithm. The appearance of the two independently moving cameras led to the system nickname of chameleon vision.

A stereoscope utilizes two pictures of the same object, taken at different angles, to produce a single image with a three-dimensional appearance. By locking two cameras onto a common target, the data relayed can provide a more accurate measure of the target's distance and azimuth from the robot. The two cameras mounted on the robot could execute one algorithm to act as a stereoscope, but could also work independently. The individual aiming of each camera onto different targets delimited each camera's field of view to the side of the robot it was mounted on. To separate the large bounding box (the area containing the visible objects) gener-

ated by two detected targets into one for each camera, a virtual window function was implemented in the program.

The optimal range of visibility of the target light was calculated considering the position of the cameras on the robot and anticipated angles to be encountered. Analysis was done to prevent the obstruction of the cameras by the arm when it maneuvered game pieces. Potential hazards and game conditions were considered to ensure the vision sensor system would perform to its potential. The target light values in the vision sensor program had to be constantly modified to account for ambient light present at different venues/playing fields.

In the event that a camera was damaged during a match, spares of all cameras and associated movement assemblies were brought to competitions. If a new camera was connected to the system, a modular program code allowed easy recalibration of the control loops using two quick set point value measurements. This assisted in quick repair of the robot (appropriately named A-Rack-Nid after the game) if needed.

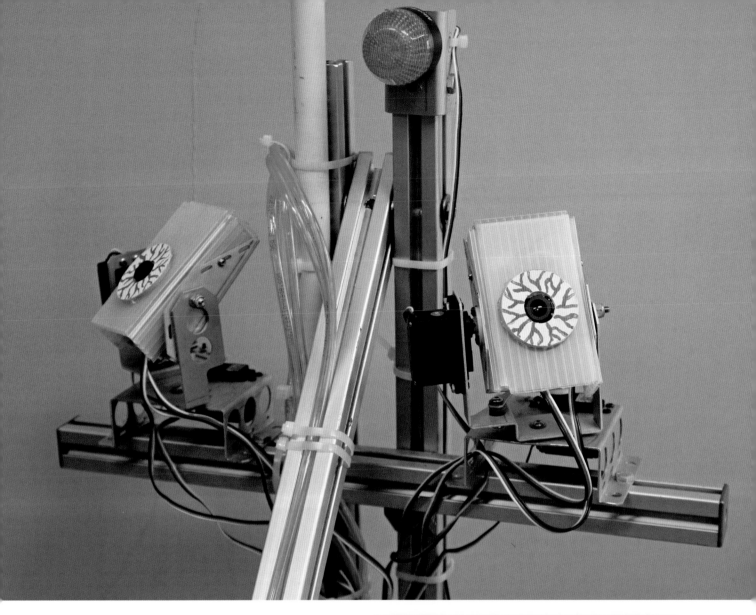

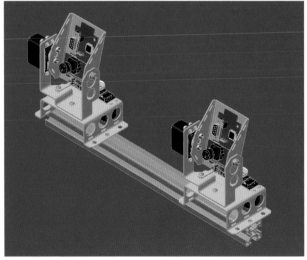

▶ A CAD drawing of the crossbar and cameras details the position and mounting of the devices with their accompanying pan and tilt controls. This comprises the vision sector of the inertial platform.

AUTONOMOUS MODE AND DRIVER FEEDBACK

A switchbox on the robot included four autonomous mode switches that, read in binary, provided a selection of sixteen different autonomous programs to choose from for each match. Three delay switches were also added to allow up to 3.5 seconds of delay, in 0.5-second increments, for the robot to defer the start of its autonomous program.

Two robot operators were required to maneuver the machine across the field. The driver had the ability to see which gear the robot was in with LEDs mounted on the operator interface. Serial data was also displayed on a laptop from a custom-written dashboard, which received and interpreted the wireless data from the robot. The positions of the camera servos, pneumatic cylinders, and status of the compressor were displayed in this way.

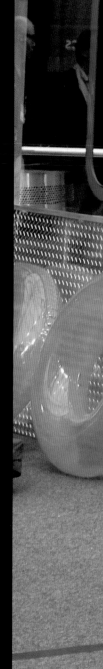

▶ A-Rack-Nid sits in starting position in its home zone, holding a keeper and waiting for the autonomous round to begin.

SCORING MECHANICS

To pick up game pieces from the floor and place them on the rack, an arm with a gripping mechanism was incorporated on the front of the robot. Two actuators controlled the two arm members, and a third actuator opened and closed the gripping fingers. To fit within the volume restrictions, the robot arm started each match folded close to the frame.

The gripper's design specifically allowed the robot to capture both inflated and deflated game pieces, with the ability to grab a game piece by one side or through the center. Textured grip surfaces were applied to both the inside and outside of the fingers.

An electronic brake prevented the robot from deviating away from scoring position. The yaw rate gyro and dual axis accelerometer provided feedback to maintain alignment with the rack.

The use of multiple motion and vision sensors proved to be the backboard of the robot's design. The inertial platform created by these sensors provided A-Rack-Nid with the necessary means to accurately and rapidly navigate the playing field and score. This small team of fewer than twenty students demonstrated their programming and integration abilities, designing a unique navigation system to master the demanding challenges of both autonomous and teleoperated competition.

▼ Three separate actuators control the motion of the two-member arm and gripping mechanism. A cable ensures that the arm does not overextend.

▶ The fingers can grip a game piece by closing on one side or by expanding in the center. This flexibility enables the robot to maneuver the game pieces, regardless of their inflation level.

◀ The robot takes off at the start of the autonomous round, locking its cameras onto a target and approaching the rack.

▲ A-Rack-Nid is designed with extra attention applied to the integration and fine-tuning of the various sensors. The result is a fully functional and successful inertial platform.

V Is for Victory

A vacuum-powered gripping device and telescoping arm were the defining features of Team 1086's robot, built and engineered by students from Glen Allen, Virginia. The robot incorporated three vacuum cups, with suction created by two venturi vacuum generators, to secure game pieces for transport to the rack for scoring. The ability to approach the spider legs from any angle to deposit a game piece equipped the team with considerable flexibility.

Before deciding on a final design, every design was evaluated using a risk analysis developed by the team. Some of the factors weighed included the ability of the device to effortlessly score on all three spider leg levels, the speed to score rapidly but efficiently, and the vulnerability of such a device to succumb to damage or defensive plays by opposing alliance members.

UNDERSTANDING A VENTURI VACUUM GENERATOR

The venturi vacuum generator supplied in the Kit of Parts was used to create a vacuum system to pick up the game pieces. The venturi effect occurs when fluid flow is constricted, as in the narrowing of a pipe or tube's diameter. This constriction causes the flow velocity to increase. As determined by the laws of nature, the pressure of the flow must decrease, thus creating a vacuum. The difference in pressure between the vacuum and surrounding atmospheric pressure, dependent on the area applied, creates the force to pick up game pieces.

▲ A wooden prototype demonstrates the effectiveness of three vacuum cups. The design was changed from an H configuration to a V to reduce weight and provide better access to scoring on the spider legs.

EVALUATING FOR MAXIMUM EFFICIENCY

The design of the vacuum system began with the evaluation of the efficiency of the venturi vacuum generator supplied in the Kit of Parts. Ideally, the system would use a moderate amount of air from the compressor. Venturi devices consume air throughout operation, and running the compressor to repressurize the vacuum circuit demanded time and battery power, two precious FIRST commodities. A strong vacuum would provide a large force for picking up game pieces, but would also consume large amounts of air. A weaker vacuum would conserve air, but not provide the force for a strong grip. The team performed tests on the use of one, two, or three vacuum generators in the vacuum circuit. The most efficient system was found to be two generators, powering three vacuum cups. This arrangement provided sufficient grip without sacrificing too much cylinder air loss. The team also experimented with different styles and sizes of vacuum cups to find the design that maintained the best grip on the game pieces. The ideal quantity of three cups was determined to sufficiently hold each piece steadily. The high-friction rubber material of each cup helped hold the game pieces firmly in place.

Team 1086 crafted a wooden prototype that mounted the vacuum cups in an H pattern, consisting of a center cup and two extension arms with a cup mounted on the end of each. One vacuum generator powered the center cup, and a second generator was connected to the two extended cups. From this prototype, proper dimensions were established to finalize a design. The H layout was modified to a V configuration, which enabled better accuracy for lining up the cups with the game piece surface, reduced the weight of the system, and made it possible for a game piece to be scored on a spider leg from a direct, head-on approach.

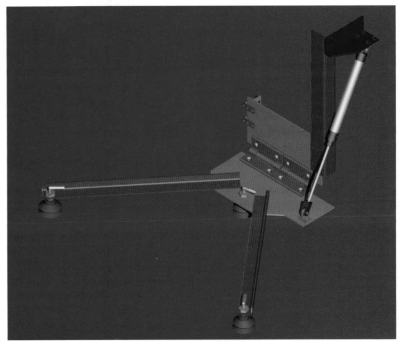

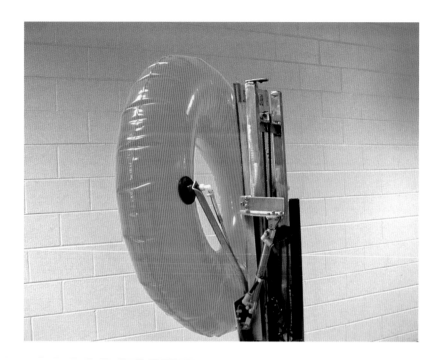

ACTUATOR OPTIONS

The position of the vacuum cup arrangement was altered between a horizontal and vertical alignment with an actuator, which tied in to the already-present compressor circuit. When extended, the actuator pushed the vacuum cup system into a horizontal position, parallel with the playing field floor and game pieces. When activated, the vacuum cups picked up a game piece. The actuator could hold the game piece horizontally, or then retract, holding the game piece vertically while the robot maneuvered to the rack. This orientation allowed the piece to be placed on a spider leg from a straight-on approach, at an angle, or over the top.

ELEVATION CONTROL

A telescoping lift conveyed the vacuum cups from floor level to the height of the top spider leg. A series of three interconnecting ladder-like rails was made from aluminum tubes and angle. The bearings of one rail were made to slide into a channel of an abutting rail.

Small rollers were fixed to the sides of the telescoping arm to prevent side-to-side movement. Such motion caused the arms to function inconsistently, but the rollers stabilized the rails and made them more efficient when lifting game pieces. A stop was added on top of the arm to keep the three rails from moving in the incorrect order and potentially sliding apart. The elevation of the rails was provided by a chain and driven by a motor coupled with a one-speed transmission.

▲ When a vacuum activates, the robot can carry a game piece in a vertical or horizontal position. An actuator positions the vacuum cup assembly either perpendicular or parallel to the playing field floor.

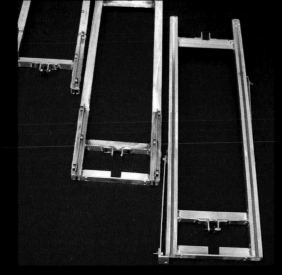

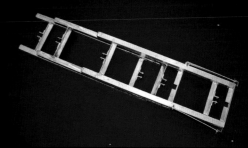

▶ Three aluminum rails are connected to alter the height of the vacuum system. A motor-driven chain system lifts each section in sequence.

▲ The retracted rails position the vacuum system low to the playing field floor. The extended actuator positions the vacuum cups evenly around the top surface of a game piece.

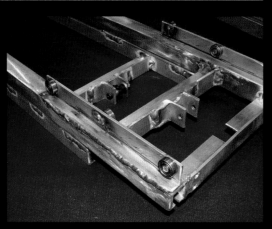

◀ Small bearings of one rail fit into channels milled out of an abutting rail. These bearings allow the parts to easily slide against each other as they are extended and rescinded.

▲ At maximum extension, the telescoping arm supports the vacuum cups at the perfect height to score on the top level of the rack.

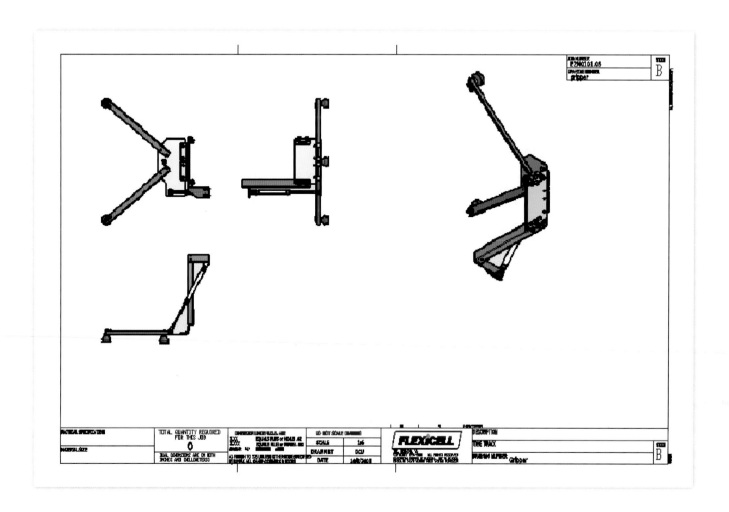

▲ A drawing of the structural components of the vacuum arrangement includes the view from different angles. This helps facilitate and expedite construction.

3D AIDING STRATEGY

Design software was an important part of Team 1086's development process. The playing field, rack, and game pieces were drawn in SolidWorks. These drawings assisted in dimensioning the required length of the telescoping arm to reach both the game pieces on the floor and the top spider leg of the rack. Measurements were computed to find the necessary size of the vacuum cup device to successfully score on the spider leg from different field approaches. Strategy for the challenge was much easier to discuss when team members had a 3-D representation of what they were approaching. Autodesk Inventor was used to draw robot compo-

nents, as well as the final assembled machine.

Careful risk analysis helped prepare the team for possible design flaws and equipped them with a clear focus on attaining speed, efficiency, flexibility, and protection. Team 1086's creativity was channeled through an indispensable understanding of venturi vacuum systems and how to obtain maximal performance. By comparing the effectiveness of multiple vacuum generators and possible vacuum cup configurations, an exemplary game piece manipulator was created. This champion machine captured the team's talent through exercise of their virtual and real-world knowledge.

▲ The Black Pearl is a result
of creative thinking, careful
analysis, and educated
design. The vacuum system
is equipped to position game
pieces anywhere on the rack.

Reliably Balancing an Alliance

▼ An alliance robot approaches Team 1985's ramp, aligning its wheels with the ramp treads and beginning its climb up the 17-degree incline.

The Robohawks, of Hazelwood, Missouri, approached the 2007 challenge with the fresh ideas and enthusiasm of a rookie team. The approach to robot design, following on the heels of the anticipated release of the new challenge, can be almost distracting with the diverse outpouring of ideas. Team 1985 focused on a clear goal: to create a robot that would be valuable to their alliance partners during competition. The team found a way to make the ever-present force of gravity work with their machine, instead of designing the robot to fight it. The result was an innovative ramp design that utilized the power of gravity and motion of alliance robots to activate the robot's defining feature.

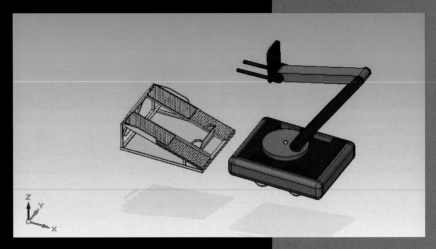

The defining feature of Team 1985's robot was the dual ramps for lifting alliance robots. A strong alliance in any match would include a machine that could reliably raise the other two robots 12" (30.5 cm) above the ground to score the thirty bonus points. The Robohawks focused their creativity on designing the necessary machinery to accomplish this.

Initial designs focused on a pneumatic lift system. Compressed air would inflate an inner tube sandwiched between two connected plates. Airflow into the tube would cause the upper plate to rise. The limited lifting power of this system, along with concerns of the weight of the system, forced a redesign.

Robot volume and weight restrictions dictated the new design. Interconnecting and hinged moving parts compiled a ramp system attached to each side.

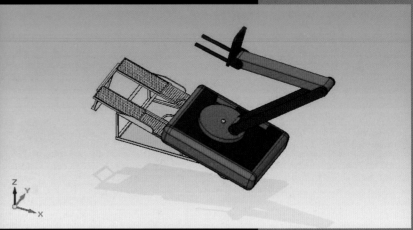

▲ As the alliance robot proceeds up the ramp, it encounters a push panel, which extends a second set of ramp treads.

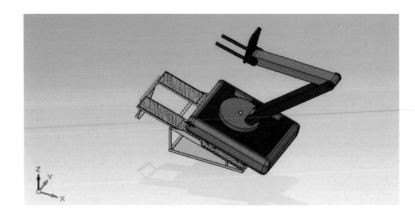

RAISING A ROBOT IN FOUR STAGES

Three main pieces comprised each ramp system. The triangular base enabled an alliance partner to drive up the ramp and onto a hinged ramp, which connected to a sliding ramp extension. The two ramp components were hinged to the top of the base and rotated around this point.

An alliance robot would first approach Team 1985's robot in the home zone at the end of a match. The wheels of the robot, with a minimum ground clearance of 2.5'' (6.4 cm), aligned with the ramp treads. The robot would begin to drive up the 17-degree incline, as its weight pushed the ramps down into the ground, preventing movement or sliding of the ramps.

As the robot drove up the ramp, it encountered a vertical push panel, which, when activated, extended a second set of ramps further up the incline. Side rails along the ramps aided in guiding the robot on its climb, and prevented it from falling off.

At full extension of the second set of ramps, the push panel folded down to allow the robot to continue its climb up the ramp. A rail placed at the top of the ramp prevented the robot from climbing too far.

As the robot climbed further up the ramps, the center of gravity of the robot-ramp system shifted. Hinges placed at the top of the first set of ramps acted as a center of rotation, and as the center of gravity passed over this point, the ramps rotated from a 17-degree incline to a horizontal position. Legs connected to the furthest end of the extended ramp contact the ground, halting the rotation so that the ramp settled at a level 12.5'' (31.8 cm) above the playing field floor.

▲ When the second set of ramp treads has been fully extended, the push panel folds down and the robot continues climbing.

▼ As the center of gravity of the robot passes over hinges placed at the top of the incline support, the two ramp sets rotate to a horizontal position. A set of back legs supports the new platform, holding the robot 12.5'' (31.8 cm) above the playing field floor.

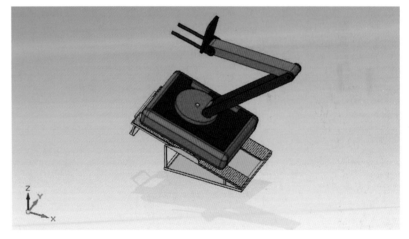

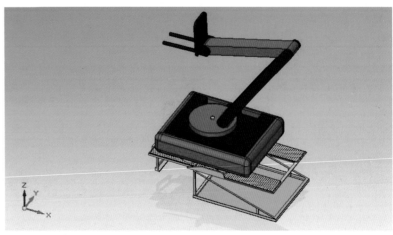

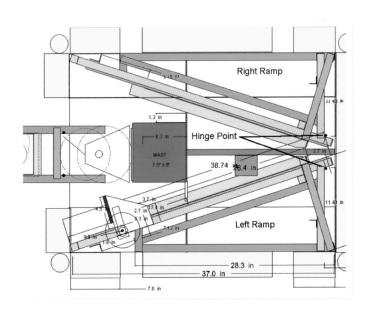

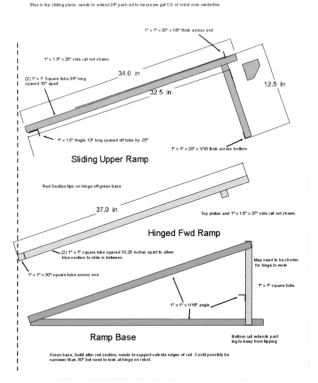

MECHANICS OF MANEUVERING THE RAMPS

The ramps were folded in position against the robot sides at the start of each match. Deployment of these ramps was required near the end of a match to engage the alliance robots. A single BaneBots motor and gearbox controlled the extension of the ramps. The motor rotated a spool that was attached to the ramps via cables and pulleys. As the cables were wound, the ramps rotated out, then down. When the center of gravity passed the center of the hinge line, gravity pulled the ramps to their final extended position.

To avoid interference between the bumpers and the ramps as they were deployed, a double hinge was designed. These hinges assisted in supporting the weight of the ramps as they were extended, and had to be durable enough to withstand the impact of the ramps hitting the playing field floor. Two strap hinges welded together, with a reinforcing steel gusset welded into the corner to provide additional stiffness, provided the required support and flexibility demanded by the complex and cumbersome ramps. Added gussets also provided support to prevent deformation of the hinges if a robot with a ground clearance of less than the designed-for 2.5'' (6.4 cm) pushed the ramps.

A bungee cord anchored from each of the deployment hinges to the opposite side of the chassis held the ramps in place during competition. A spring-loaded latch, driven by a servo motor, was attached to the top of the robot. A cable was run from the latch to each ramp to ensure that accidental deployment would not occur if the machine were to bump into one of the spider legs during competition. When the extension of the ramps was initiated by the driver, the motor released the cables and allowed the ramps to deploy freely.

▶ Three-dimensional scaled drawings of the composite ramp system helped solve potential clearance issues.

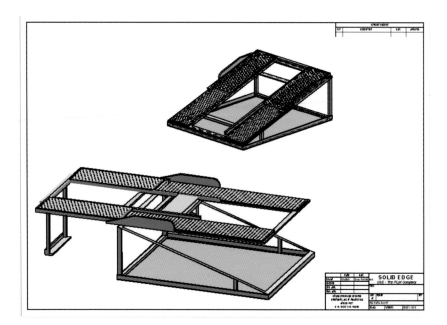

◀ A drawing of an overhead view of the ramp mechanism corresponds with an exploded side view of the ramp parts. The green base, red hinged ramp, and blue sliding upper ramp mesh to raise the alliance robots.

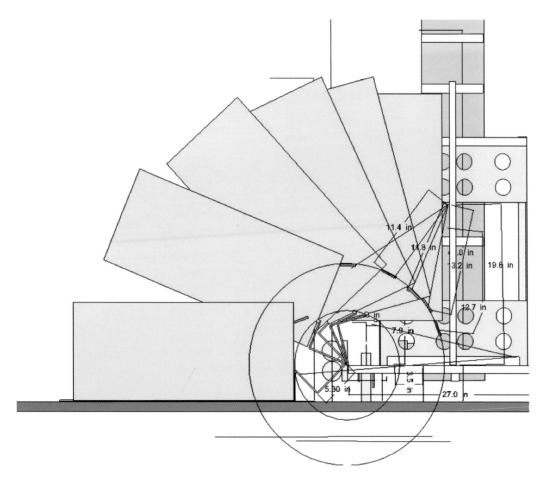

◀ The design of the double hinge allows the ramps to rotate into their deployed position while avoiding contact with the protective bumpers. A pair of strap hinges are welded together with a steel gusset for support.

In Theory's claw can grip a game piece between the outer and inner edges, holding it in a horizontal orientation. The mast raises or lowers the claw to place the game pieces on the rack. While scoring points, the ramps are folded against the side of the robot.

ADDITIONAL FEATURES COMPLETE THE ROBOT

The robot, named In Theory, (after the team's most commonly used phrase of "in theory, that's how we rack 'n' roll") had the capability to score bonus points. The ramps were, however, only one part of the robot's overall design. The machine was an integrated combination of a drive system, mast, claw, and ramps.

The drive system initially consisted of two CIM motors and a transmission on each side of the robot. When the ramp design was finalized, it weighed more than the expected amount. The team decided to use only one CIM motor and gearbox on each side of the robot. While this change sacrificed the potential power and speed of the machine, the benefit of the ramps would make up for the loss.

The mast was positioned in the center of the robot, and consisted a stationary base supported by an aluminum panel that ran from the front to the back of the robot. A globe motor drove a chain and sprocket to raise the first of a series of interconnected ladder-like pieces. A cable attached to the base of each subsequent member induced simultaneous vertical motion. Three presets programmed into the robot controller allowed the operator to quickly position the mast at the different heights of the spider leg on the rack.

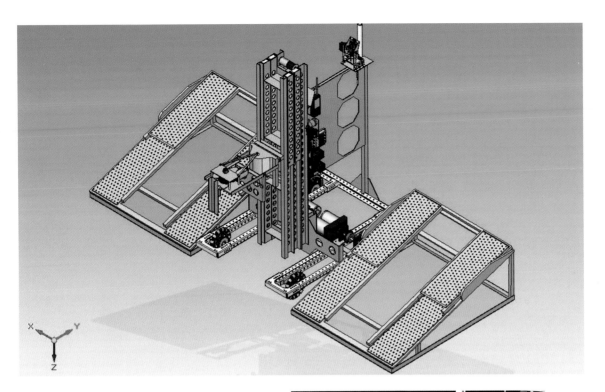

Affixed to the top section of the mast was a claw mechanism. A motor-controlled finger grasped the game pieces between the outer and inner edges, to hold them in a horizontal orientation. Limit switches provided feedback as to when the finger was fully open or closed. When the mast was at one of its three preset heights, the operator could manually raise and lower, or open and close, the claw to position game pieces on the spider leg. A feature added to the claw enabled left and right rotation if unexpected impact occurred, such as with another robot. This feature lessened the potential damage that could befall the machine.

With well-designed scoring mechanisms and reliable ramps to guarantee bonus points at the end of each match, Team 1985 fulfilled their goal of designing a robot that could best complement and assist its alliance partners. The award-winning ramps exhibited a very practical design, and used the principles of the center of gravity and pivoting in configuring the ramp layout. Rather than design a machine that fought gravity to elevate the alliance robots, the Robohawks instead manipulated gravity to do the lifting for them. Their innovative approach to the 2007 challenge has bestowed this rookie team with well-earned recognition.

◀ With ramps deployed, In Theory invites its alliance partners to board, scoring much-welcomed bonus points.

Materials Make the Robot

Innovative use of technology in FIRST robots can sometimes be seen in something as simple as the material chosen to construct the machine. The creative minds of rookie Team 1987, from Lee's Summit, Missouri, took an in-depth look at structural elements to fabricate their robot. They were awarded the Delphi Driving Tomorrow's Technology Award for their use and integration of different materials in their robot's mechanisms.

The design began with a goal of creating an arm assembly that could reach for and lift rings from the playing field floor, place them at any of the three target height levels, and deploy ramps to support alliance partners. The arm would need to be mounted on one end of the robot to create room for the ramps, and would need to be able to maneuver out of the way to allow more space for alliance robots to mount. The design progressed, and soon an arm assembly took shape that could angle up and down to adjust height level and extend or shorten to regulate reach.

A prototype of a telescoping tube, with an end-mounted vacuum-powered suction cup, was constructed from plastic. Along with a compressor and vacuum generator, the prototype was able to successfully grip and lift the inflated game pieces. This performance solidified the arm design, and a detailed engineering process began.

◄ With arm assembly raised and ramps deploying in under two seconds, the robot is ready to support two full-sized robots and score extra points.

INVESTIGATING MATERIAL PROPERTIES

Team 1987 analyzed various materials as potential tube components. Initially, chrome-molybdenum steel tubing was chosen for the telescoping tubes. Two sizes were ordered to nest together, and machining began. The team quickly realized, however, that the material, although relatively lightweight, contributed more mass than they desired. When a longitudinal slot was cut into the alloy, the tube began to spread and distort due to inherent stresses.

Repair of the tubes was determined to be impractical. The chrome-molybdenum was abandoned for use in the telescoping tubes, and further investigation into alternative materials was performed. After studying the density, strength, and structural rigidity of various other types of tubing, the team chose a carbon fiber tube.

The team conducted further calculations to determine the maximum generated torque of the telescoping system, in order to select a motor that would be able to adjust the angle of the arm without failure. BaneBots motors with different gear ratios were chosen from the Kit of Parts to power both the telescoping tubes and arm angle elevation. Software modeling was then performed to dimension parts to check for potential interferences.

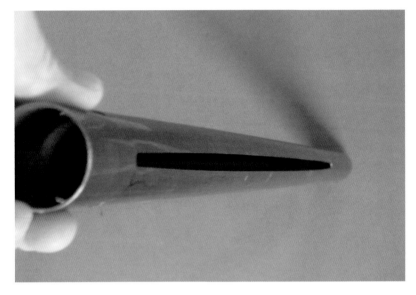

▲ Machining of the chrome-molybdenum steel tubing results in a spreading of the slot. The idea of repair is discarded, and a replacement tubing material must be found.

▼ Carbon fiber tubes of different diameters are chosen as the material for the arms, which will extend to pick up and place game pieces from the floor to the rack.

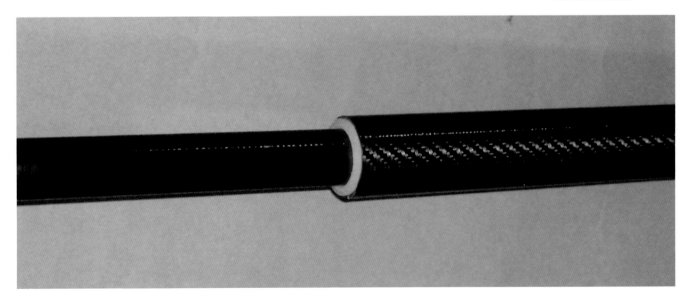

REFINING THE DESIGN

The arm assembly consisted of the telescoping carbon fiber tubes and a chain-driven powered arm pivot to adjust the arm angle, which was mounted atop a chrome-molybdenum tube on the Kit of Parts chassis. The chrome-molybdenum material was feasible as a mount, and did not exhibit the same failures as it did when it was machined or torque was applied. At the base of the chrome-molybdenum tube, two spring plungers were installed to allow the arm to twist relative to the chassis if impacted by opposing robots, protecting the device from damage.

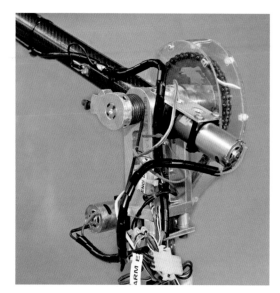

◀ The arm pivot assembly is powered by a BaneBots motor, which adjusts the angle of the carbon fiber tubes using chain. A separate BaneBots motor provides the power to telescope the tubes.

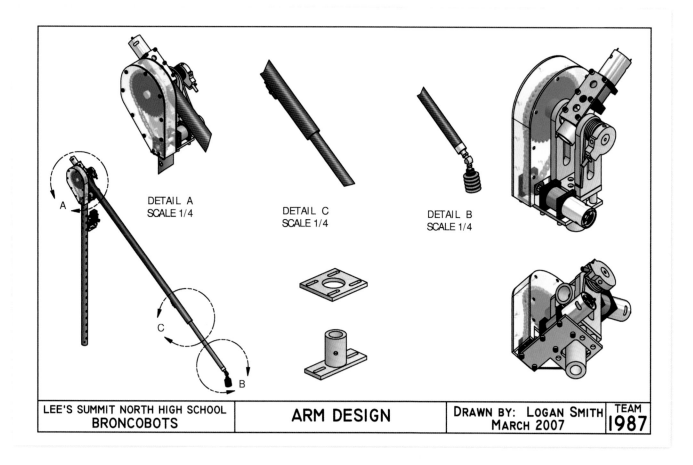

DETAIL A
SCALE 1/4

DETAIL C
SCALE 1/4

DETAIL B
SCALE 1/4

A

C

B

| LEE'S SUMMIT NORTH HIGH SCHOOL BRONCOBOTS | ARM DESIGN | DRAWN BY: LOGAN SMITH MARCH 2007 | TEAM 1987 |

The BaneBots motor that powered the arm pivot was counterbalanced to reduce the applied load from torque. A spring was wound at the pivot to preload the arm in the upward direction, thus lightening the load placed on the motor.

To enable the carbon tubes to interact as designed, a 1'' (2.54 cm) pitch lead screw was used to extend and retract the inner tube. There was very limited room to work within the tube system; the gap between the centered inner and outer tubes was only 0.19'' (4.8 mm).

A very small aluminum key was manufactured and mounted on the inside of the outer tube. This prevented the inner tube from rotating as the lead screw rotated a lightweight Kynar nut, affixed to the inner tube. The carbon fiber tubes were then precision cleaned, epoxy bonded, and heat cured to securely attach them to the pivot.

A vacuum cup was affixed to the end of the telescoping tubes to pick up the game pieces for scoring. The cup chosen was specifically made to secure soft-surfaced, curved items. A compressor and the necessary control circuitry were mounted to the robot chassis; little electrical or mechanical modifications were required of the vacuum components provided in the Kit of Parts.

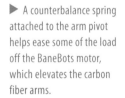 A counterbalance spring attached to the arm pivot helps ease some of the load off the BaneBots motor, which elevates the carbon fiber arms.

▶ When the BaneBots motor rotates the lead screw, it pushes or pulls a Kynar nut. This nut is attached to the inner carbon fiber tube with an aluminum sleeve. The white component is a bushing to help facilitate the motion of the tubes.

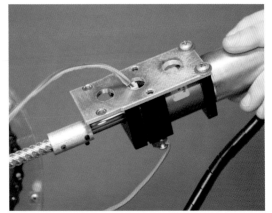

▶ The red, downward-facing cylindrical object is the location where the carbon fiber tube was bonded to the pivot with epoxy.

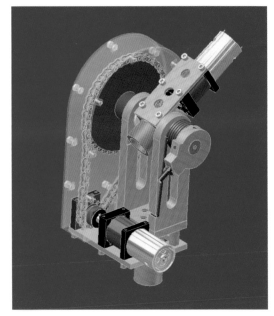

▼ Attached to the inside of the outer carbon fiber tube is an aluminum key. This key prevents the inner tube from rotating with the lead screw, so that it is restricted to linear motion with respect to the outer tube.

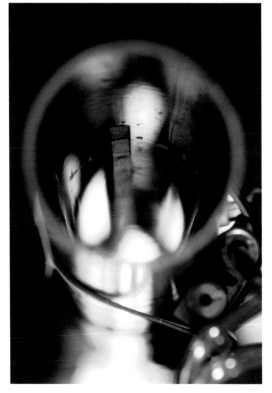

▶ Coiled urethane tubing connects the vacuum generator to the vacuum cup, seen at the end of the retracted carbon arm. This 45-degree angled vacuum cup is ideal for picking up the curved, soft-surfaced game pieces.

◀ At full extension, the carbon fiber arms reach 66" (167.6 cm). This enables the robot to pick up and place game pieces at any height, expanding its scoring capabilities.

▲ Proximity switches are screwed into the outer carbon fiber tube. Tiny rare-earth magnets bonded to the inner tube trigger the switches, preventing the arm from overextending.

INTEGRATION FOR OPTIMIZATION

To synchronize and optimize the two BaneBots motors, Team 1987 applied further gear reduction to tune the speed of each motor. Further adjustments were made to the acceleration rates of these motors with programming.

It was difficult for the driver to visually estimate the extension of the inner carbon tube, and the BaneBots motor encoders did not always start at a zero point to measure distance traveled. Additional components were needed to control the relative position of the carbon tubes to prevent hyperex-

tension. Proximity switches were attached to the outside of the outer tube, and rare earth magnets were bonded to the outside of the inner tube.

The limits of the inner tube's extension were halted when the magnets passed the switches, as a signal was sent to the robot controller. Thus, the inner tube could not extend past the outer tube, and the robot's driver had one less function to monitor. Preset positions of heights for the playing field floor and three levels of the rack were programmed, so that the driver needed only to select a preset on the joystick to pick up and place game pieces.

▶ The Broncobots' first year in competition proves to be successful. Their investigation and use of different materials helps them win the Delphi Driving Tomorrow's Technology Award.

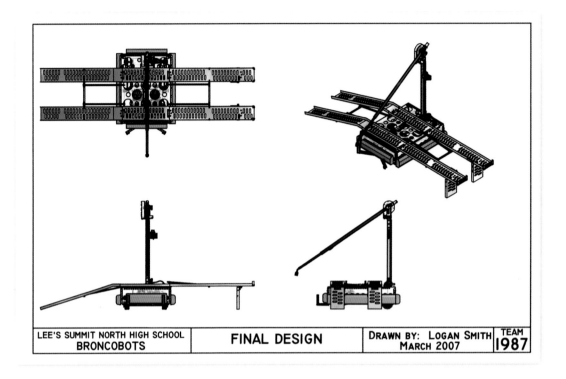

| LEE'S SUMMIT NORTH HIGH SCHOOL BRONCOBOTS | FINAL DESIGN | DRAWN BY: LOGAN SMITH MARCH 2007 | TEAM 1987 |

RETAINING ORIGINAL OBJECTIVES

A high-torque drive system and omniwheels provided power to the machine. The robot's ramps could provide protection to the arm assembly when raised, and support two full-size robots above the playing field floor when activated. This robot incorporated the simplicity and straightforwardness of the intended design, and maintained consistency and reliability. Unique materials were researched for desired properties, and then integrated to complete the robot system. The Broncobots' pride in their robot and their team is evident on their website with thorough video documentation of their design and build process. This rookie team accomplished much in their first year, fueled by an enthusiasm for learning and a passion for discovery.

General Motors Industrial Design Award

The General Motors Industrial Design Award celebrates form and function in an efficiently designed machine that effectively achieves the game challenge. This award is based on the judges' review of the team during the competition, looking at factors such as the robot's scoring ability, autonomous program effectiveness, and defensive characteristics. Winners of this award achieved superior on-field performance with well-designed robots and skilled robot operators.

General Motors is a founding sponsor of FIRST, and has been intensively involved since its inception, generously supporting events and dozens of teams in both its Michigan base and throughout the country. Senior executives have served on FIRST's board of directors. A hallmark of General Motors' involvement has been the strong mentoring culture it has nourished through the teams it supports. Professional engineers and technical staff from General Motors serve as mentors and role models for high school students, who in turn mentor middle school students to bring them the excitement and inspiration of science and technology.

Speed, Stability, Strength, Score!

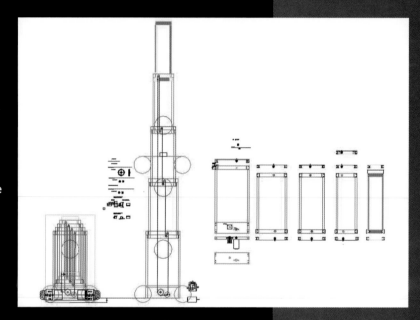

▼ A CAD drawing helps team members visualize the five stages of the mast. Their necessary sizes and interferences are fully calculated before construction begins.

An efficiently designed machine is not only aggressive on the playing field, but also reactive to the surrounding elements. Team 25, from North Brunswick, New Jersey, apportioned their efforts to build a robot that could score points while constantly responding to commotion on the field. The machine displayed an understanding of stability, and how big of a role it can play. A robot that delivered speed, strength, stability, and straightforward control such as the one built by Raider Robotix exhibited superior performance on the field.

RAPID, PREPROGRAMMED ELEVATION

For maximum scoring potential, a fast and lightweight system was needed to effectively process game pieces. Raider Robotix settled on a multi-stage forklift. The stages were built from aluminum structural tubing, and their motion was aided with sliding glide blocks of Delrin. A minibike CIM motor powered the lifting/lowering of the five sections of the mast.

The stages were lifted with a 28' (8.5 m) long timing belt, which snaked around the motor pulley, eight idler pulleys, a potentiometer pulley, and a spring-loaded idler from a CIM motor. The timing belt started and ended at the top stage of the mast, with a turnbuckle for static tension. This design accelerated the elevator to a lifting speed of 10' (3 m) per second. The mast could be controlled manually or automatically. Along with a joystick, the driver's station was equipped with a five-button control box. Each button corresponded with a different height, from ground to top of the rack. One push of a button would automatically set the height of the gripping mechanism to score on the rack. A potentiometer measured the turns of the timing belt to find the linear height of the gripper.

▶ A single timing belt runs along multiple pulleys to elevate the stages of the mast. With the help of a potentiometer, the three spider leg heights were preprogrammed for simplified and accurate scoring.

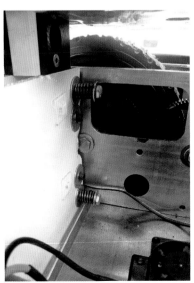

▼ Springs reinforce the frame of the mast. These allow for some flexibility in the event that there is hard contact with another robot. The mast will yield slightly to the impact, rather than snapping from rigidity.

▲ The gripper starts each match folded against the mast to fit within the starting envelope. When a game piece must be picked up or moved, the gripper folds down from a hinge, and the fingers are ready to be placed inside the center hole.

The base of the mast attached to the robot chassis with a spring-loaded mount. This mount permits some flex in the system, which protects the mast from snapping if impacted. Dual-acting shock absorbers dampen the movement and are mounted from the top of the mast base to the back corners of the robot. This spring system also helps stabilize the robot when the mast is fully extended vertically.

Gas-shock based struts were installed as supports to keep the mast from being whipped around from robot motion or opponents. These mounted from the mast to the end of the robot and reduced the sway experienced at the top of the robot.

The stages of the mast were initially lifted with standard pulleys and a timing belt. The aluminum idler pulleys were greased on the axles. The pulleys were turning too fast, and were under too large a load. The idler pulleys were replaced with permanently greased needle roller bearings, which noticeably improved the speed and power of the system.

SPEED, STABILITY, AND CENTRAL CONTROL

Team 25 agreed during brainstorming that to score efficiently, the robot would need to quickly grab, carry, and release each game piece on the rack without a heavy mechanism. These requirements were fulfilled with a device that grabbed each game piece from the center hole. The gripping mechanism folded out from the mast on a hinge when it was needed.

A globe motor mounted to the base of the gripper turned a lead screw, which pushed two pieces of square tubing away from the chassis. Two other pieces of this tubing were affixed to the gripper base, and were positioned outside the pieces attached to the lead screw. The opposite ends of the mobile and immobile pieces were connected, such that as the lead screw turned, the tubing arms spread apart.

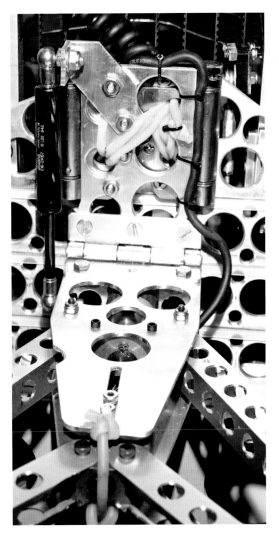

▼ The hinged base of the gripper mechanism includes a motor-driven lead screw, which spreads apart the two fingers as it is extended.

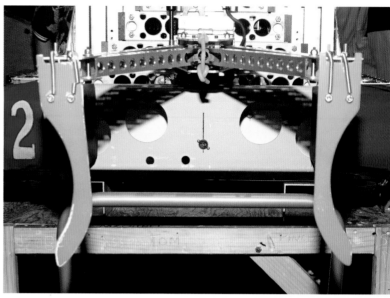

A curved finger attached to each end of the
arms with a spring-loaded hinge. When the robot
expanded from the starting envelope, the gripper folded down to a horizontal position, and the
spring-loaded fingers secured into a vertical orientation. The spreading of the arms gradually spread
apart the fingers, allowing them to grasp the inner
hole of a game piece.

This design promoted a quick game piece
capture, because the driver could start moving the
robot toward the goal as the fingers spread apart.
The orientation of the gripper held the pieces in
position and they could slide across the floor until
fully elevated.

The high speed of the lifting mechanism
required some modifications of the gripper. With
such rapid acceleration, the gripper could drop a
game piece. Sudden turning or contact with other
robots could also compromise the gripping ability.
The fingers were designed to slightly deform the
game piece when they spread, to maintain better
contact and prevent the piece from slipping. Salon
door hinges attached the gripper to the robot to
allow for flex from side to side, in the case that

the mechanism was hit. These hinges returned
the gripper back to the original position when the
motion subsided. Surgical tubing was added to the
hinge to impart tension, so that change in momentum did not cause the natural swaying reaction.

AN EIGHT-WHEELED DRIVETRAIN

The drivetrain was based on a previous competition robot, and was adapted to respond to the
2007 challenge. A long wheelbase with small-
diameter wheels would lower the center of gravity
and help prevent tipping of the robot. To achieve
this, an eight-wheel drive was created. Grooves
were machined into the treads of the wheels,
which had a relatively small width and diameter,
to further increase traction on the carpet.

Rather than building a gearbox out of aluminum plate, 4'' × 6'' (10.2 × 15.2 cm) aluminum
tubing was machined to make drive channels to
encase the components. Each channel was 37.5''
(95.3 cm) long, and contained four inline wheels
coupled by three idler gear axles. Along the inside
of the channel, the sides were equipped with a
series of six locations for input shafts. These inputs

◀ The hinge at the base of the gripper is reinforced with surgical tubing, which places the hinge in slight tension to overcome swaying motion caused by changes in momentum.

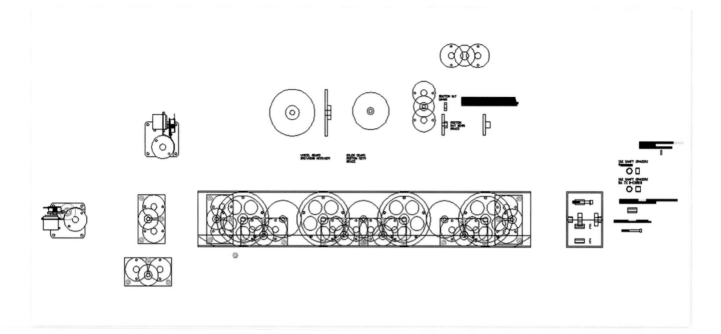

provided the option to mount the drive motor at any of these locations. This flexibility contributed to the modularity of the drive system.

Two kit-provided CIM motors were bolted to a motor plate, which meshed with a single input shaft gear. This gear then meshed with a wheel gear. A special gearbox was designed to convert Fisher-Price motors to a CIM motor face, by a gear reduction of three to one. This adaptor enabled Fisher-Price motors to be easily added to the system to provide extra power.

Servo brakes were installed with the drive system to maintain the robot's position in the event of opponents pushing. These brakes could be bolted in place of a CIM motor. The motor plates were machined so that the locking pin will protrude through a hole in the motor plate to lock the input shaft gear.

The collaboration of ideas from both students and mentors, along with comprehensive CAD documentation of robot components, reinforced the confidence of Team 25. The well-designed 2007 robot incorporated speed, power, and stability to secure a successful season. The passion and dedication of Raider Robotix prove that nothing is beyond their reach.

▲ Eight wheels of relatively small width and diameter are used to stabilize the robot chassis. Two CIM motors power the drive. The drive channels are equipped with several locations for insertion of input shafts. This provides different motor mount location options.

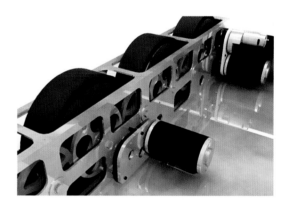

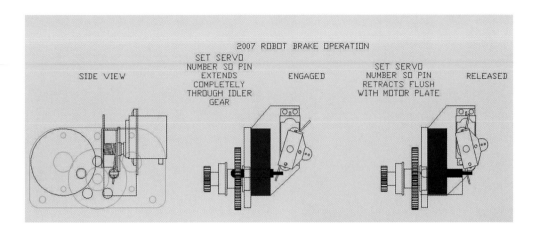

2007 ROBOT BRAKE OPERATION

SIDE VIEW

SET SERVO
NUMBER SO PIN
EXTENDS
COMPLETELY
THROUGH IDLER
GEAR

ENGAGED

SET SERVO
NUMBER SO PIN
RETRACTS FLUSH
WITH MOTOR PLATE

RELEASED

▲ Servo brakes lock the drive train in place. A servo pushes a locking pin through precast holes in the input shaft gear to halt the rotation of the gear and engage the brakes. When the servo retracts, the pin pulls out of the gear and the system is free to rotate.

▶ Thorough integration of the drive, mast, and gripper components complete the machine. Team 25's robot embodies their motto that "Nothing is beyond our reach."

Controlling the Offensive Game

Fast, maneuverable, strong, and accurate. These words describe the robot design goals set by Team 75, from Hillsborough, New Jersey. Rather than select a machine and then calculate its performance, the RoboRaiders established a goal to score at least six game pieces per match. The theoretical robot would need to travel at a top speed of 12' (3.7 m) per second to achieve this. Consequently, the machine took shape beginning with an effectual drive train and control system.

Aside from the drivetrain, the robot featured a four-axis arm mounted on a fully rotational turret. An end effector manipulated game pieces by grabbing them around the outer diameter, and was specialized to collect them from the human player over the alliance station wall. The purely offensive robot specialized in scoring, thanks to its effortless control and fluid motion.

TAKING A NEW APPROACH TO MOBILITY

The chassis was built with a 0.25" (6 mm) -thick aluminum plate, providing a rigid but lightweight frame for the drivetrain. The ideal robot would have a balance between traction and maneuverability. Omniwheels or a two-wheel drive system would make the robot very agile, but would not hold ground in a pushing contest with an opponent. A six- or eight-wheel drivetrain would have excellent pushing power, but would lack speed and fine handling. To settle for the best of both types, the RoboRaiders selected a four-wheel drive system, which also was much simpler to design than other options. Custom wheels with IFI Robotics tread in the rear and modified Skyway wheels in the front were the most efficient combination for the desired handling.

In past years, Team 75 has used chain drives to propel the drive wheels. Experience has taught the team that these chain designs are heavy, and stretching or breaking during a dynamic match is probable. They can be very high maintenance and difficult to sustain in operational order. After feeling they had exhausted all possible amendments to the chain drive design, they turned to a shaft drive as a replacement. The shaft drive provided the rugged, robust, and low-maintenance features needed to withstand a Rack 'n' Roll match.

Having never experimented with a shaft drive, a proof of principle was developed to better understand the concept and characteristics. A failure modes and effects analysis (FMEA) mapped out potential failures and all possible causes for quick recovery. A prototype was built and tested to uncover potential weaknesses.

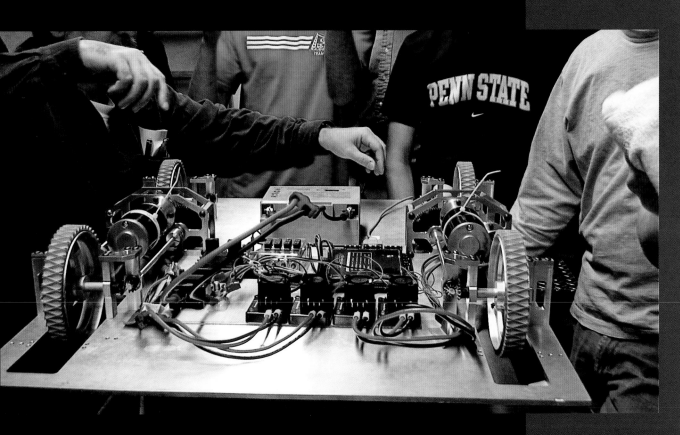

▲ The team had never built a shaft drive system before, so along with a proof of principle, a full working prototype was constructed and tested.

OPTIMIZING MOTORS, IMPROVING GEARS

The target speed of 12' (3.7 m) per second was compared with power curve graphs to determine the most efficient way of powering the drive. The resulting analysis found that the large CIM motors provided in the Kit of Parts were noticeably inefficient when utilized in this specific application. The smaller CIM motors were more economical in decreasing power consumption. From this data, a gear ratio of 9.6:1 was verified to provide sufficient handling and acceleration characteristics.

The left and right sides of the drivetrain incorporated two of the small CIM motors, coupled with the Kit of Parts transmission in an aluminum housing. Aluminum driveshafts were attached to each wheel with bevel gears, which used the output of the gearbox shafts to transfer power to each wheel with a custom right angle gearbox at each wheel.

System testing led to the discovery of large stress levels on the individual bevel gears, leading to the shearing of gear teeth. The aluminum driveshafts were too malleable and flexible, which misaligned the gear teeth under heavy weight and acceleration. The FMEA was consulted, where the error and best fix had already been recorded: different shaft material; added shaft supports; and bigger, harder bevel gears would correct the problem. The aluminum shafts were replaced with thicker titanium members, failing gears replaced with larger hardened steel bevel gears, and shaft support bearings were added to eliminate cantilevering and flexing of the shaft when under heavy loads. The problem was also corrected in the software, adding brief soft starts and stops into the controls to smooth out the shock from acceleration. The added delays reduced possible damage to the robot while improving responsiveness.

▼ A chart of primary and secondary gear reductions was used to find the optimal gear reduction between the drive motor speed and wheel speed.

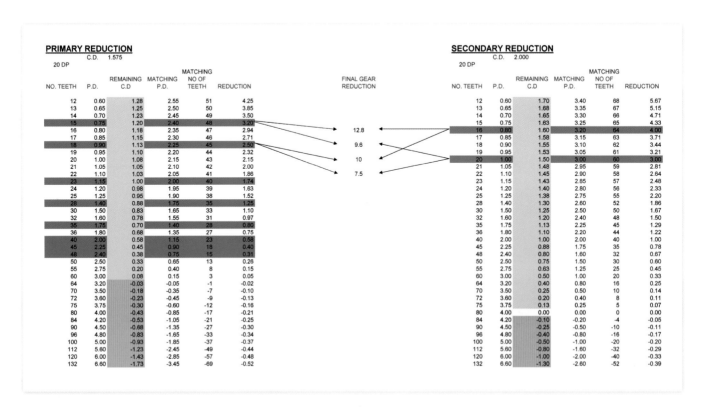

PRIMARY REDUCTION

C.D. 1.575
20 DP

NO. TEETH	P.D.	REMAINING C.D	MATCHING P.D.	MATCHING NO OF TEETH	REDUCTION
12	0.60	1.28	2.55	51	4.25
13	0.65	1.25	2.50	50	3.85
14	0.70	1.23	2.45	49	3.50
15	0.75	1.20	2.40	48	3.20
16	0.80	1.18	2.35	47	2.94
17	0.85	1.15	2.30	46	2.71
18	0.90	1.13	2.25	45	2.50
19	0.95	1.10	2.20	44	2.32
20	1.00	1.08	2.15	43	2.15
21	1.05	1.05	2.10	42	2.00
22	1.10	1.03	2.05	41	1.86
23	1.15	1.00	2.00	40	1.74
24	1.20	0.98	1.95	39	1.63
25	1.25	0.95	1.90	38	1.52
28	1.40	0.88	1.75	35	1.25
30	1.50	0.83	1.65	33	1.10
32	1.60	0.78	1.55	31	0.97
35	1.75	0.70	1.40	28	0.80
36	1.80	0.68	1.35	27	0.75
40	2.00	0.58	1.15	23	0.58
45	2.25	0.45	0.90	18	0.40
48	2.40	0.38	0.75	15	0.31
50	2.50	0.33	0.65	13	0.26
55	2.75	0.20	0.40	8	0.15
60	3.00	0.08	0.15	3	0.05
64	3.20	-0.03	-0.05	-1	-0.02
70	3.50	-0.18	-0.35	-7	-0.10
72	3.60	-0.23	-0.45	-9	-0.13
75	3.75	-0.30	-0.60	-12	-0.16
80	4.00	-0.43	-0.85	-17	-0.21
84	4.20	-0.53	-1.05	-21	-0.25
90	4.50	-0.68	-1.35	-27	-0.30
96	4.80	-0.83	-1.65	-33	-0.34
100	5.00	-0.93	-1.85	-37	-0.37
112	5.60	-1.23	-2.45	-49	-0.44
120	6.00	-1.43	-2.85	-57	-0.48
132	6.60	-1.73	-3.45	-69	-0.52

FINAL GEAR REDUCTION

12.8
9.6
10
7.5

SECONDARY REDUCTION

C.D. 2.000
20 DP

NO. TEETH	P.D.	REMAINING C.D	MATCHING P.D.	MATCHING NO OF TEETH	REDUCTION
12	0.60	1.70	3.40	68	5.67
13	0.65	1.68	3.35	67	5.15
14	0.70	1.65	3.30	66	4.71
15	0.75	1.63	3.25	65	4.33
16	0.80	1.60	3.20	64	4.00
17	0.85	1.58	3.15	63	3.71
18	0.90	1.55	3.10	62	3.44
19	0.95	1.53	3.05	61	3.21
20	1.00	1.50	3.00	60	3.00
21	1.05	1.48	2.95	59	2.81
22	1.10	1.45	2.90	58	2.64
23	1.15	1.43	2.85	57	2.48
24	1.20	1.40	2.80	56	2.33
25	1.25	1.38	2.75	55	2.20
28	1.40	1.30	2.60	52	1.86
30	1.50	1.25	2.50	50	1.67
32	1.60	1.20	2.40	48	1.50
35	1.75	1.13	2.25	45	1.29
36	1.80	1.10	2.20	44	1.22
40	2.00	1.00	2.00	40	1.00
45	2.25	0.88	1.75	35	0.78
48	2.40	0.80	1.60	32	0.67
50	2.50	0.75	1.50	30	0.60
55	2.75	0.63	1.25	25	0.45
60	3.00	0.50	1.00	20	0.33
64	3.20	0.40	0.80	16	0.25
70	3.50	0.25	0.50	10	0.14
72	3.60	0.20	0.40	8	0.11
75	3.75	0.13	0.25	5	0.07
80	4.00	0.00	0.00	0	0.00
84	4.20	-0.10	-0.20	-4	-0.05
90	4.50	-0.25	-0.50	-10	-0.11
96	4.80	-0.40	-0.80	-16	-0.17
100	5.00	-0.50	-1.00	-20	-0.20
112	5.60	-0.80	-1.60	-32	-0.29
120	6.00	-1.00	-2.00	-40	-0.33
132	6.60	-1.30	-2.60	-52	-0.39

◄ The drivetrain gearbox creates a 9.6:1 gear reduction from the two CIM motors to the wheels. There is one gearbox for each side of the robot.

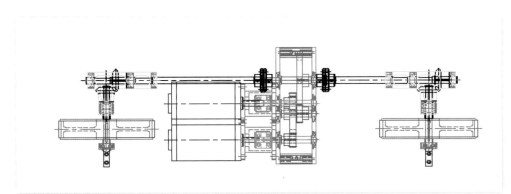

◀ The drivetrain assembly components are drawn in AutoCAD, and a three-dimensional model is made with Inventor. Every detail of the drive system is documented to ensure correct interaction between components.

► Testing of the shaft drive resulted in gear teeth breaking off due to unacceptably high stresses in the drive system. The shaft and gears were replaced, and shaft supports were added to correct this problem.

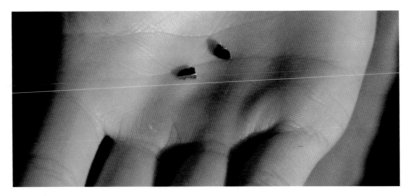

▼ The rear wheels are covered in IFI treads to provide the traction and pushing power needed on the playing field. Two 45-degree beveled gears connect the wheel to the driveshaft.

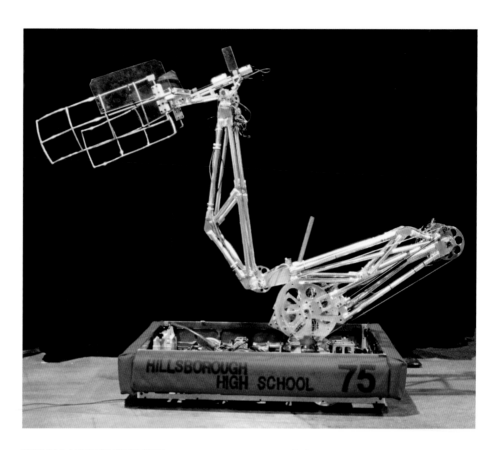

IT'S ALL UNDER CONTROL

The intricate control system eased the interaction between driver and machine. A customized feedback loop was developed to better control the multiple functions on the robot. Integration of the operator interface and robot controls lessened the need for required driver inputs to control the robot. Subroutine macros were used to stimulate preestablished configurations. The driver could press a button on the operator interface to enable the macros, embedded in the electrically erasable programmable read-only memory (EEPROM). Bits of code can be erased and rewritten repeatedly from a small chip on the robot controller. The programming allows the drivers to simply push a button to position the robot in a certain pose and record data of the arrangement. The macros enable new positions to be recorded without requiring a laptop computer to be connected to the system. This feature allows the driver to record the position of the robot when it is about to score at a certain level of the rack, and recall this position later during a match.

▲ Preset positions were programmed to facilitate the scoring of game pieces on the rack. With a simple push of a button, the robot driver can position the arm to the lowest spider leg, the middle leg, or the highest spider leg.

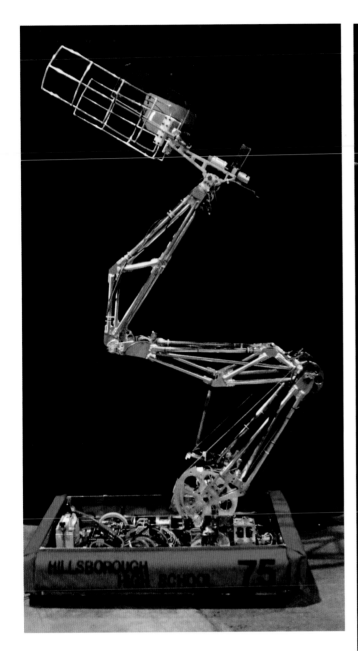

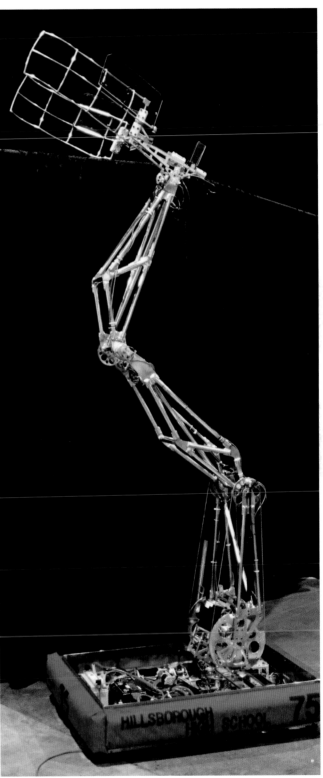

LEDs on the operator interface were calibrated to maintain accurate tuning of the joystick trim fluctuation without use of a computer. Potentiometers and encoders provided constant data regarding motors and position to the feedback loop. Warning LEDs illuminated if the arm approached the maximum size boundary or if the turret over-rotated.

The speed of the robot was monitored with code and a proportional damping algorithm minimized shock on the drivetrain due to rapid changes in speed or direction. Ten preset positions were recorded for the arm with potentiometer data of where the arm was and where it needed to be. Four of the settings were for loading game pieces, and six were for placing them on the rack. The driver controls incorporated buttons for each pre-set routine. The driver could override all motions if needed.

The thorough programming and attention to detail optimized the efficiency of Team 75's robot. The approach to an unfamiliar design using proof of principle and FMEA shows a very organized team, determined to overcome any challenges they meet. The programming applied to the various components eased the burden on the robot operators and increased the efficiency of all functions. Such a machine has well earned the General Motors Industrial Design Award by notably exhibiting both form and function in its overall design.

▶ A control panel with multiple buttons and switches is wired to the joystick inputs. The driver can use this panel to send the robot into preset positions, and LEDs inform the driver of system conditions.

▶ Potentiometers are used on the joints of robot components to save and recall preset positions as well as define the coded limits. Encoders are placed directly in line with driveshafts to count wheel revolutions.

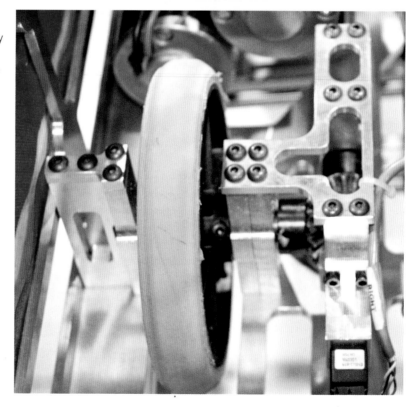

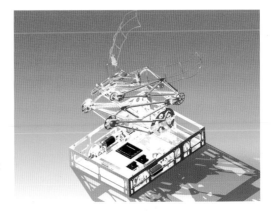

▶ The robot is folded up to fit within the starting envelope at the beginning of a match. Careful attention to design detail produces a drawing that is almost identical to the final product.

Designing an Anthropomorphic Arm

The highest priority for Motorola's Team 111 was its scoring system. The team favored an arm with many degrees of freedom to score on any level and have the dexterity needed for defense. A swerve-drive system, where each of the four drive wheels could be independently steered, complemented the arm with an ability to out-maneuver opponents and make fine adjustments while scoring.

The resulting arm was a complex system that required significant design to incorporate all of the driving mechanisms within a tight design envelope. The robot design started with a kinematic analysis to understand the required robot geometry. Hand sketches identified the arrangement and integration of components, and detailed CAD plans evaluated the fit and function of each part. Autodesk Inventor was an important design tool to plan the layout of all components and produce detailed plans to manufacture components.

Team 111's robotic arm was modeled after a human arm. The final design included a shoulder joint, an upper arm, elbow joint, forearm, and fingers. The shoulder rotated from the front to the back of the robot and the upper arm rotated 180 degrees at the elbow. The upper arm and forearm pivoted 270 degrees along the forearm's axis. The gripper, mounted at the end of the forearm, consisted of two sets of three fingers. Deviating a bit from the human model, the upper arm extended 10" (25.4 cm) to reach the upper rack for scoring.

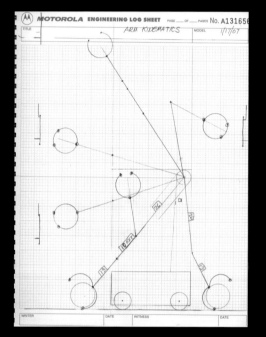

◀ Team 111's very complex arm began with a sketch to analyze the system's kinematic motion. The sketch illustrates that two points of location and an extending segment are needed to gather tubes at both ends of the robot and score them on all levels of the goal.

▼ Accurate Autocad Inventor CAD plans are created to design all components of this complex system. Skilled use of the CAD software allows the design to be evaluated early on during the design process.

▶ The arm rotates at its shoulder and elbow. The arm also pivots at the shoulder to allow the forearm and claw to rotate for scoring tubes when the robot is not directly in front of the goal.

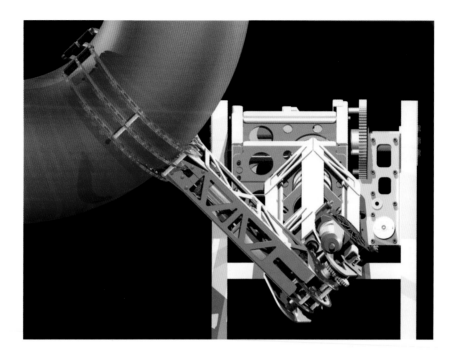

DESIGNING AN ARM IN SIX WEEKS

Not only did the arm have to be highly maneuverable to manipulate tubes, but it also had to be strong enough to resist damage from other robots while protecting the internal components. The kinematic sketches specified the geometry for each arm linkage. This drawing also identified the necessity to extend the upper arm for retrieving tubes at the front of the robot and for scoring at the top level. Further review of the system revealed the need to rotate the arm normal to the shoulder's axis to enable scoring a ringer from the side in addition to straight ahead.

With such a complex system planned, the biggest challenge the team faced was time. The arm required the majority of the team's design effort. To speed the design process, three mechanical engineers divided up the detailed mechanical design of the arm. The three engineers directed teams that designed the shoulder, upper arm, and forearm. Each system was modeled with Autodesk Inventor software. Assemblies of systems were first joined in the computer. Because separate teams developed each component, it was critical that the components were first assembled in a master file to confirm their compatibility before parts were fabricated. Using Autodesk Inventor significantly cut down on the amount of design errors in such a complex electromechanical system.

Autodesk Inventor also helped speed up the design process by perfecting mechanisms. For example, testing revealed that the original gripping system could hold a tube, but the tube rotated when the arm rapidly moved. The dimensions of the pivot points for the pneumatic pistons that powered each finger were manipulated in the Autodesk Inventor drawings. With the CAD software, the pivot points were manipulated to increase the resulting range of motion. Conducting these tests with the software model saved significant time. Once the location of the computer model was verified, the results were applied to the physical system to improve performance.

▲ Every component of the robot is modeled and integrated with each other using Autodesk Inventor design software. In addition to serving as a platform to determine the location of all components, the software provides engineering drawings to manufacture parts.

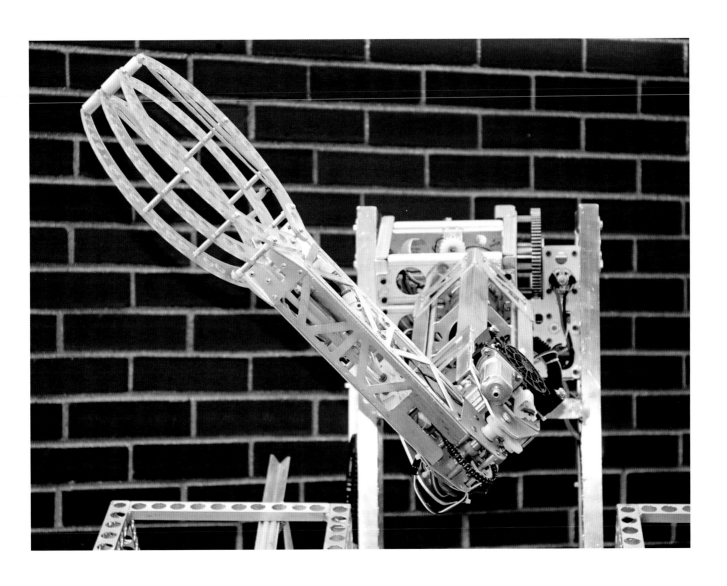

SHOULDER JOINT DESIGN DETAILS

In addition to the mechanical framework, each joint also required that the power system be integrated into the design. Each motion required a separate power source and gearbox, with the complete motor and transmission integrated into the joint. In addition, each joint was monitored with sensors to report the position to the operator interface, and each joint was a conduit for conductors leading to other arm components. Including all these components in a compact package was challenging.

The shoulder joint was the most complex mechanism as it allowed rotation in two planes: along the shoulder's axis and along the bicep's axis to twist the upper and forearm. Worm-gears were used with both of these motors, since these gears held the arm's position when power was not supplied to the motors. Motors were selected not only for their ability to deliver the needed power, but also for their size.

The small Fisher-Price motor provided in the kit was ideal as the shoulder's power source. The original plans called for a gearbox that reduced the speed of the motor by a factor of 421 times. This motor's small size allowed it and its gearbox to be mounted on the fixed tower that connected the arm to the robot base. The gearbox was designed with an option to change the internal gearing to speed up or slow down the shoulder's rotational speed. This option was exercised when the robot was tested and the shoulder rotated at a speed faster than could be easily controlled. To regain control, and gain additional lifting force, the gearbox was outfitted with additional gears to slow down the shoulder.

The thin profile of the supplied Keyang motor made this motor the ideal choice to rotate the forearm and upper arm. Equipped with a worm-drive transmission, this motor's output was strong enough to supply the needed torque and the correct speed for controlling the motor. With its slim profile, the motor was mounted along the shoulder's axis.

▶ The initial plan for the shoulder joint is detailed in a pencil sketch that specified the location and size of each gear in the worm-drive transmission. The combination of gears shifts the rotational angle by 90 degrees and reduces the output speed by a factor of 421.

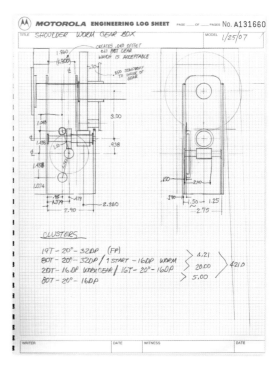

▼ The rotational gearbox is attached to the partially fabricated shoulder joint. The motor, used to twist the upper and forearm, is mounted along the shoulder joint.

◀ Every inch of real estate is used in the completed shoulder joint. In addition to housing two motors and their transmissions, the shoulder joint also includes power and control wires, pneumatic lines, sensors, and mechanical stiffeners.

▲ A frame extends from the shoulder joint as a bearing to support the arm's twisting motion. The smaller upper arm is attached to the shoulder at the end of this frame, with the even smaller forearm attached to the upper arm.

▶ The arm's twisting ability provides the dexterity needed to place tubes from any given orientation between the robot base and the goal.

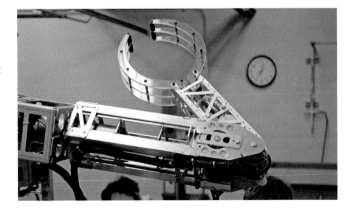

MINIMIZING WEIGHT AT THE ARM'S END

It was essential to minimize weight at the end of the arm, as any load here created torque that had to be counteracted by the shoulder motors. The Denso window motor was the ideal motor to use at the elbow because of its lightweight, built-in worm-gear box. As with the shoulder motors, this type of gearbox locks in position and does not back drive when power is not supplied to the motor.

To deliver linear motion at an economical weight, pneumatic power was used in the arm extremities. A single piston extended the upper arm to facilitate scoring on the upper level. The forearm consisted of a set of nested frames, with the inner frame fixed to the shoulder joint and the outer frame supporting the forearm. The piston connected the two frames, thereby extending the reach of the forearm and gripper when expanded.

The forearm supported two pistons, each with a 1'' (2.54 cm) stroke length, that powered each side of the claw. Incorporating the elbow sprocket, pistons, solenoids, and fulcrum components into the forearm was another significant design challenge aided by the use of Autodesk Inventor. Minimizing weight was a design priority at this extremity, and all components were selected and the parts were manufactured to meet the minimal weight criteria.

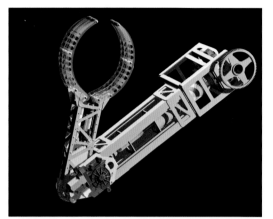

◀ All elements of the arm are apparent in the CAD drawing and robot photograph. The arm rotates at the shoulder, twists at the shoulder, rotates at the elbow, extends along its upper arm, and includes a claw at the end of the forearm.

▶ A pneumatic cylinder provides power to extend the arm's reach for scoring at the top goal level. The arm's bicep is a nested set of frames, with the inner frame sliding out when the piston is extended.

▶ The forearm is a compact system that houses two pneumatic pistons to open and close the claw. The forearm's frame is very light, yet is still able to protect the components and provide sufficient strength to support the claw and its load.

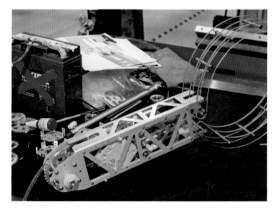

MEETING DESIGN GOALS

The team's emphasis on designing a fully capable arm meant that other design features had to be sacrificed. For example, the software control team focused its work on making the arm controllable, and did not devote much design time to the robot's autonomous functions. Also, a ramp was added to the robot to give the team an alternate scoring system, but this bonus feature was a lower priority to the primary means of scoring. The team's ability to keep its emphasis on a primary design goal was key to designing a complicated system in a short period of time.

The human arm was the inspiration for Team 111's design. Recognizing the constraints of time and resources, Team 111 minimized the robot arm's degrees of freedom while achieving their design strategy. Even with this reduced number of degrees of freedom, the design was extremely complex and a highly efficient precision system that earned the respect and admiration of many other teams.

▶ Team Wildstang catches the attention of all—not only for their colorful look but also for the wildly effective humanlike robotic arm.

Simply the Best

The potential bonus points that could be earned by alliance robot elevation motivated many teams to focus their design on ramps, arms, or other elevation devices. Team 148, of Greenville, Texas, anticipated this, and decided to build a purely offensive robot. Each alliance needed only one robot with a lifting device to score the points, leaving two other robots to carry out different game strategies. The Robowranglers were confident that a reliable scoring robot was essential to a successful alliance team. They focused their energies on a specialized design that concentrated on the purely offensive strategy of scoring and descoring game pieces on the rack.

FULL CONTROL WITH ROLLERS

The interface between machine and game piece was a combination of a claw and roller system, which was prototyped using Vex Robotics components to determine functionality. The Vex parts enabled full-sized models to be rapidly constructed for testing. These models were fundamental in defining the dimensions of the components that would result in efficient interaction with the game pieces. When the team became satisfied with prototype testing, the claw and roller components were drawn in detail using SolidWorks.

▲ Vex Robotics components make great full-size prototypes. The interaction of these prototypes with the actual game pieces helps the team prove the feasibility of claw designs.

▶ The driving force behind the claw is the fingered upper roller, which quickly pulls game pieces into the claw. Two BaneBots motors and polycords impart opposite rotation to the rollers.

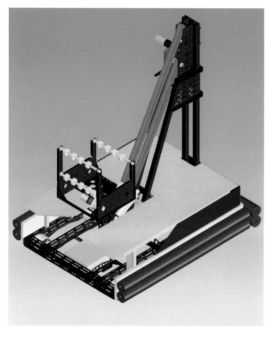

◀ Every component of Team 148's robot is carefully modeled in SolidWorks after the prototype testing is concluded. Modeling helps with construction and complete archiving of the 2007 machine.

◀ An early version of the claw has only polycord running over pulleys to contact the game pieces. The longer extension of the lower rollers allows the robot to pick up pieces in orientations other than horizontal on the floor.

Dual rollers, spinning in opposite directions to lure a game piece into the claw frame, made contact with each game piece in such a way that regardless of opponents' defensive actions, the robot maintained full control over the piece. A single BaneBots motor, connected by round polycord belts and delrin pulleys, drove each roller. These connecting materials efficiently transferred the torque of the motors to the rollers while remaining lightweight and flexible enough to slightly yield if a game piece was encountered that was not fully aligned with the claw.

Both rollers were mounted parallel to the ground, in front of the robot, so that when collecting a game piece, the claw floated just above the playing field floor. The lower rollers of the claw extended forward slightly past the upper rollers, enabling the robot to collect game pieces propped up against the opposite wall of the playing field, and remove opponent's spoilers from the rack. The upper roller was initially designed with only polycord running over several parallel pulleys to contact a game piece, but this did not provide enough grip on the pieces, causing them to be pushed across the floor rather than collected.

The upper roller was replaced with PVC pipe, which had holes drilled into it with inserted flexible tentacles. These small fingers were tested to establish the ideal size and quantity to most effectively grab a game piece as the roller rotated, and snugly pull it into the claw.

BENEFITS OF THE CLAW

The 12" (30.5 cm)-wide claw corralled each game piece so that it could not rotate or tilt out of position once inside. By gripping each piece in the same way, the robot could score from a predictable orientation every time. Because the claw could only handle one game piece at a time (complying with competition rules), the claw could approach a pile of pieces and only collect one, whereas a finger-and-thumb gripping mechanism would have difficulty in selecting a single piece. The reliable grasp easily accepted game pieces fed over the alliance station wall by the human player. The design of the robot physically prevented the claw from breaking the plane between playing field and alliance station, preventing possible penalty or disabling of the machine.

The frame of the claw consisted of a rigid and strong yet lightweight reinforced sheet metal shell. The bottom surface of the claw incorporated a thin Lexan shield, which protected the belts that powered the lower roller and relieved friction as the claw slid across the floor.

◀ The claw is in position to collect a game piece. A compressor and pneumatic tanks are mounted to the chassis to enable the extension and retraction of the actuators that position the wrist.

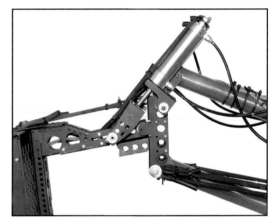

▶ To lock the wrist, a strip of surgical tubing deploys a piece of sheet metal when the wrist is first activated in a match. The lock limits the maximum extension of the wrist to 45 degrees, rather than vertical, for ideal scoring on the rack.

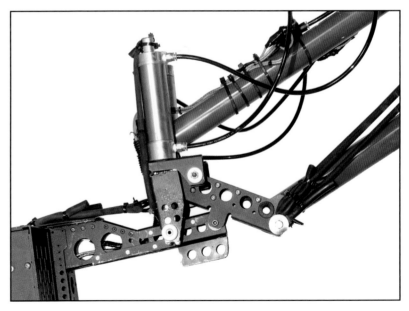

▲ The robot approaches the rack with a game piece held at a 45-degree angle. Once the top of the piece is over the edge of the arm, the wrist snaps down to firmly place the piece on the arm.

THREE-POSITION LOCKING WRIST

The Robowranglers needed to fit the claw mechanism within the starting envelope size, while maintaining the ability to reach the top spider leg. The most efficient way to accomplish this was to position the claw in a starting vertical position. Once a match started, the claw would have to turn 90 degrees to a horizontal orientation to obtain game pieces. To score, the claw would approach the rack at a 45-degree angle. A custom joint was designed to achieve these three necessary positions.

The starting position of the claw was the only time throughout each match that it was vertical. Once a round started, an actuator shifted the claw into a horizontal orientation. The team determined that it was easier to score with a game piece positioned at 45 degrees to the spider leg rather than vertically. The tip of a piece was positioned over the end of the arm, and the wrist snapped down to firmly place the game piece on the arm. A lock was made for the wrist to prevent it from extending past 45 degrees to return to a vertical position once play started.

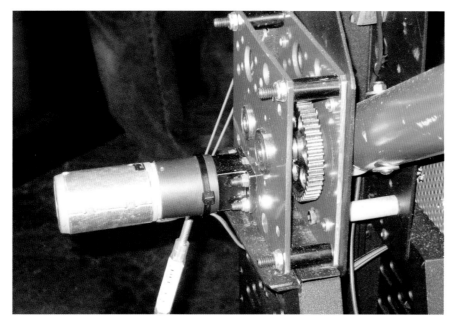

◀ The shoulder joint controls the motion of the four-bar linkage, which raises and lowers the claw and wrist. A globe motor and single-speed transmission are modified to couple with the end of the lower PVC arm.

FOUR-BAR EXTENSION

To score on all three rack levels and reach over defensive robots, a four-bar linkage arm and shoulder joint extended the claw and wrist. A four-bar linkage consists of four rigid members attached by pivots to form a closed loop. This linkage translates circular motion produced by motors into linear lifting motion. The long arms of the Robowranglers four-bar were made from 1.25'' (3.2 cm) PVC pipe, which was lightweight and provided enough flexibility to prevent damage from lifting. Three of the four system pivots consisted of a 0.5'' (1.3 cm) outer diameter aluminum shaft with a 0.25'' (6 mm) bolt through the center. The PVC arms were fitted with aluminum inserts for support and low-friction bronze iolite bushings in which the aluminum shafts rotated.

The fourth pivot was the main joint of the arm, which connected the lower four-bar arm to the robot chassis via a shoulder joint. A globe motor and transmission were modified and attached to couple the output shaft with the end of the lower arm. The globe motor was selected because it is not heavy, provides moderate power to lift the lightweight game pieces, and runs at a relatively low speed, so minimal reduction is needed.

The weight of the arm assembly was neutralized with the addition of a weight balancer. Elastic tubing stretched from the shoulder joint diagonally across the four-bar linkage. The most balancing support was provided at the highest load state, greatly reducing the load on the motor. In the position to reach the claw to maximum height, the two PVC arms were both vertical. The four-bar was mounted atop a sheet metal tower, and optical encoders on the arm provided position feedback to the drivers.

When the arm was fully lowered, the claw was positioned at the perfect height to score on the lowest of the spider legs. Once the claw obtained a game piece, the wrist joint was raised and the piece placed on the arm, without moving the four-bar linkage arm. The directness of this move helped the team score five game pieces in thirty seconds at a regional competition, and an alliance high score of 291 points at the Championship.

Team 148 focused their design efforts on a robot that could effectively complete the challenge of game piece scoring. The claw mechanism needed only to touch a game piece for the robot to command control of that piece until successfully scoring it on the rack. Assembly of an impressive and efficient vehicle achieved the goals of the Robowranglers. The integration of these simple yet elegant systems created a machine that was much more than the sum of its parts.

▶ The four-bar linkage translates rotational motion of the motors into linear motion to lift the claw. The design of the linkage system also holds the claw parallel to the floor when the wrist is extended down.

▲ At maximum horizontal extension, the four-bar arms are parallel to the floor. The claw is extended past the chassis to score on the middle spider leg. Surgical tubing eases the load on the globe motor by balancing the weight of the arm.

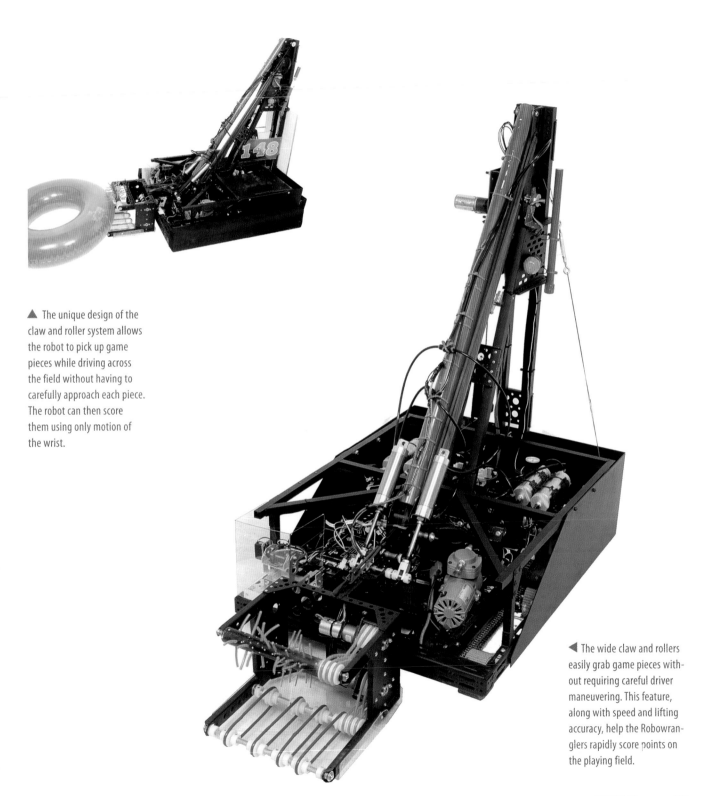

▲ The unique design of the claw and roller system allows the robot to pick up game pieces while driving across the field without having to carefully approach each piece. The robot can then score them using only motion of the wrist.

◀ The wide claw and rollers easily grab game pieces without requiring careful driver maneuvering. This feature, along with speed and lifting accuracy, help the Robowranglers rapidly score points on the playing field.

A Winning Design Process

Team 330 from Hermosa Beach, California, is a notable team for two unique reasons. First, the team is comprised entirely of home-schooled students. The students and their parents, many of whom are engineers, gather as a community to participate in FIRST. The second distinctive attribute of their team is the list of accomplishments for the season: two regional competition victories, a regional Motorola Quality Award, a regional design award, and the Championship General Motors Industrial Design Award. Part of their success stems from veteran mentors, some of whom have been with the team for ten years. The team also includes three team alumni, two of whom have since earned their own college degrees in engineering.

Team Beach Bot was successful in their endeavors because of their structured approach to engineering design. As a small team, labor was a limited resource. The robot was constructed with components that were assembled quickly, simply, and without extensive training. Plywood was a favorite building material because of its strength and its ease to be worked on with standard tools such as a jigsaw. The simplicity of the design was paired with the highest degrees of reliability: a powerful combination for on-field performance.

▲ Experience and adherence to a detailed design process is a powerful combination that fuels Team 330's performance as a fierce competitor. The robot's high degree of reliability is credited to its simple design.

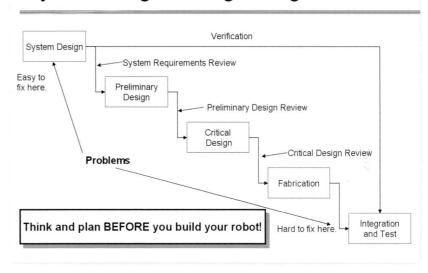

System Engineering Design Process

System Design

Verification

Easy to fix here.

System Requirements Review

Preliminary Design

Preliminary Design Review

Critical Design

Critical Design Review

Problems

Fabrication

Think and plan BEFORE you build your robot!   Hard to fix here.   Integration and Test

APPLYING A SYSTEMS ENGINEERING DESIGN PROCESS

As a FIRST veteran, Team 330 understands the need to make informed decisions at each stage of the design process. Borrowing techniques from the mentor's professional careers at NASA's Jet Propulsion Laboratory, Raytheon, and Northrop Grumman, the team implemented a systems engineering design process to create their robot. The process begins by ensuring the team's underlying strategy is sound, followed by the design of a robot to execute the strategy.

A miniature goal became a platform to quickly evaluate possible strategies. Simulated play highlighted that there was no single strategy that would always win the game. An effective defensive maneuver was discovered during this process: by scattering as few as six tubes around the rack, the opposition could be limited to rows of three tubes. With only three tubes in a row, the score would be kept low enough to make the end-game bonus points the deciding point.

Once the game was understood, a System Requirements Review was conducted and identified sixty-six design requirements. These requirements included the ability to score on all levels of the goal, the need to be fast and strong, and the use of a central repository for all revisions to the computer code. Because the requirements were written down, the team was able to verify compliance with each requirement throughout the entire build period.

The premise of a systems engineering design process is that problems are easy to fix at the earliest stages of design and more difficult to correct in advanced stages. To examine the developing design, a Preliminary Design Review was conducted once the robot was first designed. A Critical Design Review examined the operation of each subsystem when they were constructed to confirm the feasibility of the design. Only after these reviews were actual components for the robot fabricated.

▲ The Systems Engineering Design Process consists of a series of steps that advance ideas into solutions. Evaluation, testing, and improvements are the important components of this sequential process.

▲ A small-scale model of the competition provides a quick method to evaluate the game and determine possible strategies. The simple game model correlates with the simplicity of the team's robot: a characteristic of Team 330's approach to effective design.

REACHING ALL LEVELS WITH A PIVOTING ARM

The team's emphasis on reliability and limited machining capabilities motivated the adoption of a pivoting arm to gather and place tubes. With a single joint, the number of moving parts on the arm was minimized, thereby meeting a team design criteria for dependable performance.

The strategy review created a plan to quickly score a ringer on the middle level of the opponent's side of the goal. The plan also called for the next ringers to be placed on the far side of the goal. This aggressive offense drew defense from the opposition and left the alliance partners free to score on the side closest to the drivers. The ringers also became a visual barrier for the opposing alliance when trying to score on the far side of the rack.

This strategy was important for designing the arm, for it specified the ability to score on the middle level as the primary purpose of the arm. As such, the arm was designed to be most effective in this orientation. Nearly perpendicular to the floor, the arm had a long enough reach to score even when an opposing robot was between the Beach Bot and the rack. The system requirements also specified the need to score on all three levels of the rack: a task that was achieved with the single pivot arm.

After an initial design on paper, a mock-up of the system was created. Two pieces of wood connected with a drawer slide between them served as a prototype system to evaluate the utility of the simple design. A foot was attached to the extending end of the arm. When extended, this foot would slide under a tube. When retracted, the foot would hold the tube against a fixed restraining device on the other piece of wood, with the tube thereby pinched between the two attachments.

The final configuration of the arm consisted of a tube within a tube. The outside fiberglass tube was fixed to the pivot point at its end and was the primary structural member. A piston was located inside the outer tube and connected to an aluminum tube that moved in and out as the piston extended and retracted. A Lexan plate was mounted to the end of the inner tube to grab game pieces such that when the piston retracted, the game piece was lifted against a gripper and held in place. The arm's simplicity greatly contributed to its reliability and robustness.

▶ An accurate sketch illustrates the feasibility of using a single-pivot arm to score on all three levels of the rack. Embedded in the drawing is the fixed orientation of the tube and the necessity for this orientation for scoring.

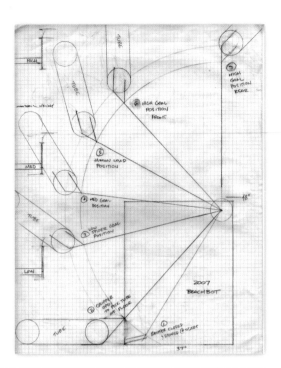

▼ The value of careful planning is apparent in the orientation of the tube as it is placed on the middle row. The tube angle closely matches that in the sketch, and the reach of the arm allows the team to score while the robot base is positioned a safe distance from the rack.

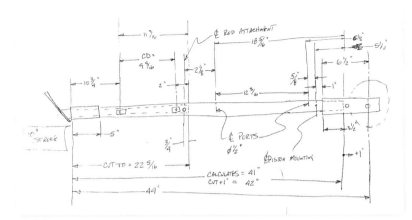

◀ Hand sketches identify critical measurements for the arm. All attachment points are located and the required stroke lengths are specified in the original sketch.

▼ CAD drawings illustrate the external frame and internal mechanism of the arm. A piston mounted inside the outer tube extends the inner tube to retrieve game pieces.

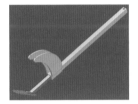

THE BEACH BOT WAVE

The human hand served as the model for Team 330's manipulator. The telescoping paddle, referred to as the thumb, pulled tubes into a fixed restraint to grab tubes. Designing the end effector to pick up tubes between a thumb and a fixed restraint was not as easy as originally thought. The perfect angle of the fixed restraint had to be determined to maximize the surface area that would be gripped.

Starting from a careful sketch, four different sets of gripping systems were evaluated. The final design resembled a wave: an appropriate shape for the Beach Bot. Aircraft-grade plywood was used to fabricate the two side plates of the wave. Rubber tread was affixed to the foot and the wave's outer tip to maintain better control of a captured tube.

All components of the gripper system were optimized during extensive tests associated with the Critical Design Review. For example, the foot length was designed to be long enough to grip the tube but short enough to avoid catching on the goal when tubes were released. Also, the gripper had to grasp the tube tightly to prevent it from falling out if Beach Bot collided with another robot, but the gripper also had to easily release the tube during scoring maneuvers. By the time a practice frame and drivetrain were assembled, a near-final version of the gripper was ready and the team could practice its ability to score.

Like the arm, the gripper's effectiveness resulted from detailed design, extensive testing, and simplicity. In the extended position, the foot slid under a tube. When the piston retracted, the

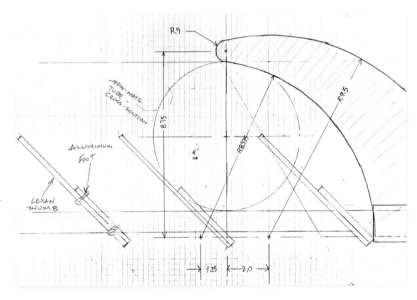

▲ Tube diameter was a critical dimension to design the sliding and fixed segments of the gripping device. The foot extended to retrieve tubes and retracted to squeeze the tubes against the wave-shaped retainer.

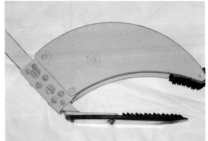

◀ The gripper was constructed from high-grade plywood, carefully cut to match the design template. Rubber tread, aligned with edges to maximize grip, is added to the edges of the mechanism.

tube was tightly pinched between the foot and the wave. This firm grip was essential to maintain the tube's orientation as it was placed on the rack, as originally sketched during the preliminary design phase. The correct tube orientation was critical to enable the arm to score the tube on any level.

PRACTICE MAKES PERFECT

The final stage of the systems engineering design process was integration and testing. To rigorously test their design, Team 330 first built a practice robot. This robot platform solidified the tube-scoring process and perfected each component on the robot. After the practice robot was built and improved, construction of the competition robot commenced.

By building the practice robot first, the team was able to make several improvements, including changing the location of a cross member in the frame to better accommodate the gripper. In addition, the practice robot provided a platform for the team to use once the competition robot was shipped. By honing their skills with the practice robot, the drivers were well trained and proficient at tube retrieval and scoring. Undoubtedly, the large amount of practice before the regional and Championship competitions greatly contributed to the team's high levels of performance.

The primary advantage of the systems engineering design process was the ability of Team 330 to know exactly what it wanted before building the robot. The value of the design process, stemming from years of experience, was validated during competition where Team 330 excelled in the areas of performance, quality, and creative design.

▲ In the extended position, the foot slides under a tube. When retracted, the tube is lifted into the mouth of the gripper to securely hold it in place.

▼ Beach Bot's ramps provided another means to score. The team's design criteria required that the ramps be capable of being lowered and raised during competition to avoid penalties if inadvertently deployed during a match.

Collaboration Sparks Success

Team 1126 thrived as an organization that promoted collaboration, communication, and valuing each other's input. Team SparX's robot was unique in that it was completely student built. While mentors guided the process and provided valuable insights, the robot was designed and constructed by the team's student members. The students modified their initial hopes of designing a robot capable of every imaginable feature to construct a very capable machine that quickly acquired tubes and placed them on the spider legs to score.

Demonstrating the team's openness to ideas, some of the principle systems on the robot resulted from conferring with the high school's custodian, who demonstrated how the school's forklift operated. The custodian also showed team members the winch system for raising and lowering the basketball backboard. This meeting's relevance was apparent in the final robot that included a winch-driven, forkliftlike elevator mechanism.

An extendable arm was attached to the elevator, with a rotating wrist connecting the tube manipulator and the arm. The creative design only required two motors, one for the elevator and one for the wrist. Two pneumatic pistons extended the arm's reach and a third opened and closed the manipulator. The resulting robot was simple and effective.

► Before any construction began, every part on Team 1126's robot was drawn using CAD—a design skill that saved time and money. The computer model verified that each part aligned with the main robot frame and provided a means to estimate the weight of each component.

◄ A winch winds a cable to lift the inner mast and elevator. The rotating wrist positions the tube at the ideal angle for scoring on a spider leg.

▼ The winch cable is routed through the outer and inner masts before connecting to the elevator carriage. The elevator carriage supports the robot's manipulator arm and the camera.

▲ An Autodesk Inventor drawing of the winch system models each component and specifies the dimensions of the completed system. The winch powers a cable that was connected to the top of the elevator carriage.

GRIP AND LIFT

The tube manipulator system consisted of a gripping mechanism that was mounted on an elevator. A winch lifted and lowered an elevator carriage along a two-stage mast. A serpentine cable was rigged through the lift such that the single cable controlled the height of the gripper and inner section of the mast. The gear ratio on the winch was modified to provide the power needed to lift the mast at an acceptable rate.

The most important feature on the robot was the gripper. After evaluating prototypes, the chosen gripper held tubes at three points along their outside edge. By doing so, the center of the tube was kept open to maximize scoring potential. A piston connected the gripper's two arms and its fast reaction time provided a quick method for grabbing and releasing tubes. A switch was located on the gripper's rear pin to detect when a tube was within the gripper's reach. When activated, the gripper automatically closed—a design feature that improved performance when the operator's view was blocked.

Other design details enhanced the gripper's effectiveness. Precise and coordinated movement of the gripper's outer arms resulted from a pair of meshed gears attached to the arms' pivot points. This transmission produced the identical amount of motion in each arm to accurately grab and release tubes.

The gripper was mounted on a wrist that was capable of rotating the mechanism through a 180-degree range of motion. This allowed the gripper to be stored within the robot's starting envelope and unfold at the beginning of each match. The wrist also aided scoring by placing each tube in the best orientation for placement on a spider leg.

PERFECTING PROGRAMMING DESIGN

The team's design expertise was also demonstrated by its programming proficiency. The entire control system was designed as modules that isolated each subsystem from each other. This approach was chosen so that changes to one part of the program were less likely to impact other aspects of the robot's software.

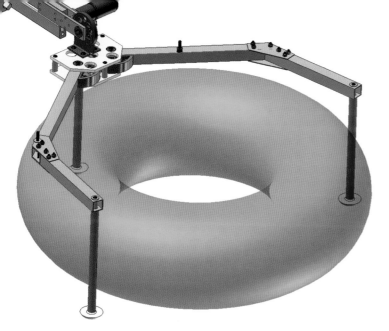

◀ Three points of contact support the tube along the outer edges and keep the center open for scoring tubes on the rack.

▶ Dual pistons extend the arm's length, with each piston enabling a specific reach to be achieved. Counting the default position when both pistons are retracted, three distinct arm lengths can be achieved with the piston extenders.

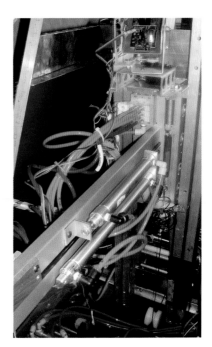

▼ A gearbox reduces the lifting speed of the wrist motor while increasing its torque. A rotary potentiometer records the wrist's rotation.

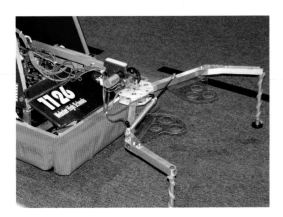

▲ A single piston connects the manipulator's arms. Meshing gears at their pivot point force each arm to move at the same rate and distance as its partner.

In addition to modular design, the control software also included team-written subroutines that improved the robot's performance and made it easier to write new programming algorithms. Computer instructions were written to buffer transmitted and received data within the program itself, using memory locations in the robot controller. The ability to store and recall data was especially useful while diagnosing and debugging the robot's camera. As another example, an interrupt-driven analog scanner function was written in the code to enable the program to run as fast as possible while acquiring data from sensors such as the heading-sensing gyro.

The team created a linear interpolation subroutine to estimate the value of any function specified as a table of data. This capability was used to estimate trigonometric functions associated with the distance to the goal and to convert voltage-based potentiometer measurements to angles and distances. Other software tools created by Team 1126 included a clock module to determine wheel speed and to monitor the time left in the autonomous period. Generic control algorithms were also written that included the capability of tuning the controller for each application.

AUTOMATING TASKS FOR IMPROVED PERFORMANCE

Hardware and software systems were integrated to automate robot functions and improve the robot's performance. The team's scripted program that was executed during the autonomous periods was one example of such automation. Like many teams, Team 1126 used the camera to locate the goal, but unlike many teams, they did not use the camera to directly drive the robot.

Team 1126's autonomous sequence began with automatically moving the robot forward to a position where the camera could see the target light. Once the signal was acquired, the data was analyzed to determine the distance to and angle from the target. This directional information was used to steer the robot to the goal using a gyro for measuring the robot's heading and wheel encoders to determine the distance traveled. After the robot moved to the commanded position the camera was reactivated to verify the robot's position. When in the proper position, the captured tube was automatically placed on the rack.

The goal-locating program was also available during the teleoperated portions of the competition. A button on the driver's station signaled the robot to find the target light and score a tube on the rack. This function was useful if the operators encountered a visual impairment when the robot was at the far end of the field or behind the rack.

A four-speed automatic transmission system on the drive wheels was another example of Team 1126's expertise integrating the robot's software and hardware. The four-speed transmission required frequent shifting based on specific speeds to optimize the power delivered to the drive wheels. To free the driver from having to manually shift the robot at each specified speed, computer code was created to monitor the average speed and automatically shift gears.

During testing, it was noticed that the transmission could not be shifted if the robot was operated under a high load, such as when turning. To reduce the load on the transmission, the automatic transmission program momentarily halted power to the drive motors for a fraction of a second. During this brief period of a reduced load, the pneumatic shifter was energized to successfully change gears. Automating this process freed the driver from monitoring speed and shifting, and allowed the driver to concentrate on the game dynamics.

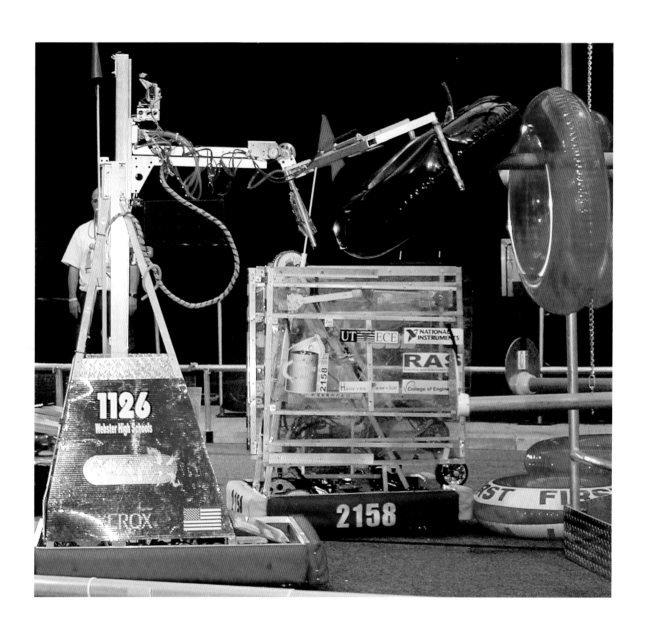

◄ With its fast scoring ability Team 1126's student-built robot was a valued alliance partner. In addition to knowing how to build a robot, the team knew how to play the game and use spoiler tubes to knock down the opponent's high scores.

A TOP TEAM CREATES A TOP ROBOT

One of the more challenging aspects of Team 1126's design process was selecting mechanisms from a variety of alternatives. The team relied on prototyping to explore design solutions and they created many unique and plausible solutions for each robot function. After experimenting with each design the team met to select the best option for each function. As each system was designed and constructed, it was evaluated to determine if the completed system met the design specifications. When problems occurred, such as the inability of the original gear ratios on the winch to lift the elevator, the entire team analyzed the problem to find solutions.

The unified hardware and software in Team 1126's robot allowed many robot functions to be automated, enabling more effective control of the robot. Coordinating the activities of the subteams that designed the drive platform, manipulator, and elevators was essential to construct a high-performing robot. Constant communication and collaboration, combined with an environment that valued each team member's opinion, became important factors to not only build a high-performing machine, but also important to create a high-performing team.

Motorola Quality Award

The Motorola Quality Award celebrates machine robustness in concept and fabrication. Winners of this award have strong, well-built robots that require little maintenance. Although the award criteria do not include on-field performance, winners of this award are typically among the highest ranked teams at the competition. As durable and reliable machines, the Motorola Quality Award–winning robots are effective competitors and valued partners.

Motorola is a founding sponsor of FIRST and has been extraordinarily generous. As a long-term supporter of FIRST teams and events, the company's commitment to workforce development has helped to inspire a whole new generation of scientists and engineers. The company has also funded independent evaluations of FIRST programs, and helped rebuild the FIRST Lego League program in Louisiana and Mississippi in the aftermath of Hurricane Katrina. Top-level Motorola executives have served on FIRST's board of directors for many years, and a retired chairman of the Motorola board currently serves as an honorary director of FIRST.

Evaluating Options with Prototypes

Without a machine shop readily available, Team 100 faced a considerable obstacle while designing a robot for the 2007 competition. Faced with such a limitation, the team designed a robot that required a minimal number of machined parts while using as many off-the-shelf components as possible.

One example of the success initiated by a limited ability to fabricate parts was their highly reliable drivetrain, which was constructed from commercially available components. The drivetrain consisted of pneumatic wheels, two-speed commercial gearboxes, $^3/_8$" (9.5 mm) bolts for axles, pillow-block bearings, steel sprockets, chain, and four motors from the FIRST Kit of Parts. No machining was required to integrate the components, and the parts were assembled using handtools.

This emphasis on simplicity aided the team's goal of building a robust robot while following an aggressive design schedule. The schedule forced design decisions to be made early based on the performance of prototypes. Once a general mode of operation was identified for each subsystem, detailed tests were conducted to evaluate the many options in each subsystem. The tests were important not only to evaluate options, but also as a means to initiate innovations that improved performance.

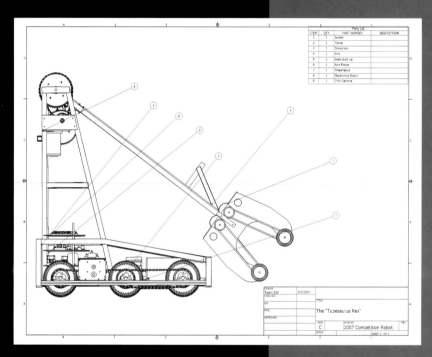

▲ A two-speed transmission powers three drive wheels on each side of Team 100's robot. A rotating arm supports a large claw that includes an upper and lower pair of rotating belts to grab and hold tubes.

▲ A three-dimensional rendering of the completed robot accurately models the physical device, even to the point of matching graphics for the team name and the DreamWorks Animation sponsor's logo.

PROTOTYPES LEAD TO PERFECTION

The importance of prototyping as a design evaluation tool was illustrated by the metamorphosis of the claw. Five potential designs were first evaluated using Autodesk Inventor CAD software and by constructing prototypes. The operating principles for the examined systems included variations of a pneumatically driven mechanism, a vacuum suction cup, and a belt-driven retrieval system.

The claw's prototyping process revealed many unseen problems. For example, while the pneumatic mechanism was found to be easily manufactured and extremely effective, its size caused concern. As a large appendage protruding from the front of the robot, it was susceptible to damage from other robots. Its size also proved to be too unwieldy and difficult to maneuver. Other designs, such as a scoop and pincher, proved to be less effective than had first appeared on paper.

Prototyping eventually led to the final configuration of the gripping system. Named the sucker, the manipulator used belts connected to an outer pulley to pull tubes into its grip. Since each belt was independently controlled, the position of the tube could be adjusted in the claw's grip. More refined testing determined the proper tension and alignment of the belts, with the belts positioned to enable them to be compressed as a tube was drawn into the manipulator. To keep the belts centered in the pulleys' grooves, belt guards were added to the manipulator—an alteration that required external rubber tires to be added to the outside of each pulley to grip the tubes.

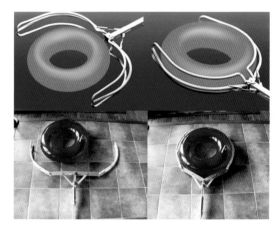

▶ An early version of the claw is modeled using CAD and constructed to evaluate the used of a single pneumatic piston to retract the arms. Testing revealed that the device was too bulky and fragile for use in competition.

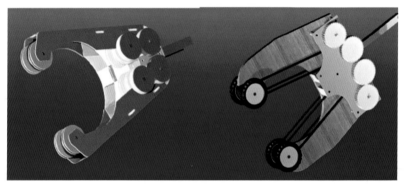

▲ Autodesk Inventor is used to design options for the robot's tube manipulator. The three-dimensional models help determine the shape and arrangement of components on the manipulator.

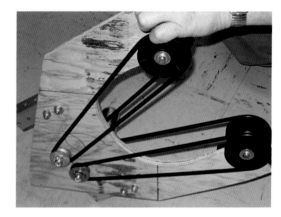

▶ Prototypes progress from computer models to physical systems. The wide spacing between the drive pulleys and the free-spinning pulleys at the claw's mouth tension the belt as a tube is drawn in.

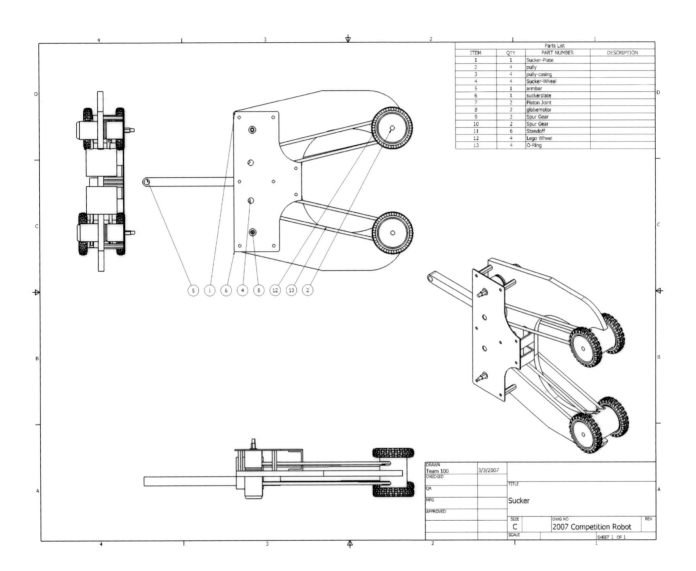

Parts List			
ITEM	QTY	PART NUMBER	DESCRIPTION
1	1	Sucker-Plate	
2	4	pully	
3	4	pully-casing	
4	4	Sucker-Wheel	
5	1	armbar	
6	1	suckerplate	
7	2	Piston Joint	
8	2	globemotor	
9	2	Spur Gear	
10	2	Spur Gear	
11	6	Standoff	
12	4	Lego Wheel	
13	4	O-Ring	

DRAWN				
Team 100	3/3/2007			
CHECKED				
QA		TITLE		
MFG		Sucker		
APPROVED				
		SIZE	DWG NO	REV
		C	2007 Competition Robot	
		SCALE	SHEET 1 OF 1	

▲ The final design of the claw includes wheels mounted at the claw's mouth. Because the drive belts are contained at the extremities inside PVC shields that keep the belt on the pulley, wheels are added to establish a positive grip on tubes.

Similar refinements were made to the other major systems on the robot based on the performance of prototypes. The prototype drive system and robot base was assembled in just five days and their performance validated the team's decision to power the robot with off-the-shelf components. Heavy duty ANSI #25 roller chain was used instead of lighter alternatives to increase the system's robustness as the heavy chain required less precision for its alignment and was less prone to breaking under severe loads.

The drive system prototype also provided a platform for evaluating other robot functions. A tower and wooden arm were added to examine the robot's ability to collect and score tubes. The arm needed to be strong enough to lift the 10 lb. (4.5 kg) claw, and support the mechanism that rotated the arm and claw. A wooden arm was used as a prototype to examine the loads placed on this part of the robot.

Testing revealed that the arm needed to be very strong to withstand collisions from other robots. Based on the prototype, the final arm was constructed from box aluminum welded together as a truss. The point of the arm supported the claw and the wide base of the arm distributed all loads across a wide area to minimize the stress at any single location.

EXPERIMENTAL TESTS TO DETERMINE DETAILS

Detailed tests were conducted to refine the prototypes and determine design details. A variety of materials were examined for the claw. While the first prototypes were constructed from plywood, variations for this system were created using molded ABS plastic and by fabricating a metal frame. Although the plastic was very light and strong, it was too flexible to keep the belts and pulleys aligned. The number of mounts for the motors, gears, and pulleys required a large area, and

the metal version of this design proved to be too heavy and cumbersome.

Based on these tests, the team returned to using wood for the final design. To evaluate the strength of various materials, samples of each material were tested. A pneumatic piston was used to load each sample piece, with a pressure regulator varying the applied load. The pressure in the piston, measured with an attached pressure gauge, was slowly raised until the sample cracked.

Of the five tested materials, fiberglass-coated plywood was the strongest. Marine-grade nine-ply Baltic-birch plywood was ultimately selected due to its high strength-to-weight ratio. The wood was reinforced with three layers of fiberglass to provide flexibility and avoid flexing. For additional strength, carbon fiber rods were embedded in the fiberglass at locations of high stress. The resulting robust manipulator allowed the arm to be left in a prone position as it had the strength to resist damage from other robots bumping into it.

Tests on the arm produced similar improvements to the design. A worm-gear transmission was required to keep the arm extended without constantly supplying power to a motor. The weight of the arm and manipulator, coupled by the arm's long length, demanded a large motor to provide the high amount of torque. Testing revealed that a kit-provided motor with an integrated worm-gear transmission was not suitable for this high load as the internal gears were made of plastic.

A custom gearbox was designed and constructed for this application, and augmented with a gas piston to provide additional lifting force. Constructed from 0.25" (6 mm) aluminum plate, hardened shafts, and bronze and stainless steel gears, the gearbox was strong enough that a single transmission and motor could supply the needed force to lift the arm and manipulator. The transmission design was also simple enough to be constructed by the team despite their limited machining capability.

▲ A completed prototype is used to test out the feasibility of the design. The arm design begins with wood as a quick method to lift the manipulator.

▶ The final arm is fabricated using aluminum in a triangular configuration to distribute forces imparted on the manipulator along the largest possible area at the connecting joint. The open area in the robot base is required to store the manipulator at the beginning of each match.

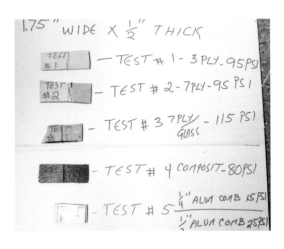

1.75" WIDE X ½" THICK

TEST #1 — TEST # 1 - 3 PLY-.95 PSI

TEST #2 — TEST # 2 - 7 PLY - 95 PSI

TEST #3 — TEST # 3 7 PLY - 115 PSI
GLASS

— TEST # 4 COMPOSIT - 80 PSI

— TEST # 5 ¼" ALUM COMB 15 PSI
½" ALUM COMB 25 PSI

◀ To choose the material for the claw, a load test is applied to various materials such as fiberglass-covered plywood, aluminum, and composite materials. The test apparatus is built with components from the Kit of Parts.

▶ A worm-gear transmission is designed and constructed to provide the force for lifting the arm. A worm-gear is needed to develop the needed torque and to keep the arm in position when power is not supplied to the arm's motor.

CONSTRAINTS PROMOTE INNOVATIVE SOLUTIONS

Size limitations spawned innovative thinking. The drive system and open center, needed to store the claw, left little room in the robot base for the electronics. To fit all the electronics in this small area, they were stacked in four layers. Each wire was labeled on both ends to identify their connection port and purpose to ease maintenance and troubleshooting. By keeping the electronics within the frame of the base, these systems were fully protected from damage.

To store the claw within the robot's base, a passive wrist was designed. Adding a wrist motor to the arm created an additional load that would reduce the manipulator's effectiveness. Instead, a simple spring-powered joint was constructed to allow the claw to collapse back on the arm when stored. Surgical tubing served as a spring that snapped the claw into position once the arm was raised. This passive system provided an additional benefit as it aided the team's scoring ability. The inherent flexibility allowed the manipulator to flex when tubes were placed on the spider legs and prevented the robot from being entangled in the rack.

The team's goal of developing the base and drive system with commercial components also produced an innovative design feature. The base was constructed from structural aluminum frame along its sides welded to aluminum channel along the front and rear of the robot. The aluminum structural pieces provided a convenient means to mount the transmissions and pillow blocks.

Slots were cut into the frame to slide bolts into the frame's channel to secure the components. The rails provided the capability to adjust the location of the pillow blocks and served as simple devices to adjust the tension on the drivetrains. As with the passive wrist, the simplicity of the mounting method increased the reliability and usefulness of the system.

TUBE-A-SAURUS-REX

Team 100 developed a computer animation of the robot to demonstrate its capabilities. Since the nondescript robot seemed too inanimate, the programmer added eyes to the claw and turned the claw into a head. This prompted another student to add some detail to the actual claw, where upon the eyes and teeth were painted on. In the end, the face developed as the robot's principle feature, which led to naming the robot Tube-a-saurus-Rex. Like its prehistoric ancestor, the modern mechanical rex was a fierce competitor due to its strength and agility.

▲ The claw pivots at the end of the arm, with surgical tubing used as a spring to align the claw with the arm once the arm is moved from its stored position. As a passive system, the wrist is a simple and effective solution that met design constraints.

▲ The final robot, named Tube-a-saurus-Rex, has a long reach and is capable of scoring on all levels.

TEAM 190

Innovatively Competitive

The noise from 15,000 fans watching the final match of the 2007 FIRST Robotics Competition was deafening as two robots climbed on top of a third. With relative ease each robot climbed gradual ramps that sprouted from the host robot. As the match ended, a single control button was actuated, pistons fired, and the gradual ramps were lifted 12 inches (30.5 cm) off the playing surface floor to secure sixty points and win the 2007 FIRST Robotics Competition World Championship.

The smooth execution of this final strategy was the culminating success for FIRST Team 190. Just as athletes dream about catching the winning pass to win the Super Bowl, FIRST Team 190 began their season with dreams of lifting two robots at the buzzer to become the 2007 FIRST World Champions. That dream was born the very first day of the season when the team adopted a strategy to be innovatively competitive.

Team 190, sponsored by Worcester Polytechnic Institute, combined their traditional philosophy of creating innovative machines with the requirement that all parts of the robot be highly competitive. The team has a long history of designing and constructing ingenious mechanisms, but the designs were sometimes prone to failure due to the complexity of the components. For the 2007 season, Team 190 adopted a design philosophy that the innovative design features of their robot needed to be robust enough to withstand the demands of intense competition. Being the top team out of over 1,300 competitors, Team 190 proved the value of their foresight.

▲ The design process included brainstorming, a formal Preliminary Design Review, continual subteam interaction, and a Critical Design Review to examine all aspects of the robot. CAD models were created to integrate all aspects of the robot and meet the size and weight design constraints.

DESIGN PROCESS ESSENTIALS

In addition to their strategy to be innovatively competitive, Worcester Polytechnic Institute's Team 190 was guided by a careful and methodical design process to generate ideas, review concepts, and create a winning robot. The team began the season with a set of prioritized goals: to score offensively by lifting two robots, collecting tubes from the floor and placing tubes on the rack, and by scoring a tube in the autonomous period of the competition. Lifting two robots was the highest priority since it was predicted that the equivalent method to achieve sixty points—placing six tubes in a row on the rack—would be a difficult task. It was predicted that an effective set of ramps would be able to win most matches.

After groups brainstormed to design robot components that accomplished these goals, a decision matrix was used to evaluate the designs. Criteria such as low design risk, speed, agility, power consumption, and robustness were evaluated to select the optimum design attributes. The decision matrix assisted the team in deciding that the robot would have a six-wheel drive, pneumatically actuated tube manipulator, telescoping rack, and dual ramps that were also pneumatically actuated.

The specific components were designed by teams of ten students, with each team taking on an individual part. All of the components were modeled using CAD software, and the parts were integrated to verify fit and function. In addition, the computer model served as a tool to estimate size and weight.

Once the parts were modeled in CAD, a formal Preliminary Design Review (PDR) was conducted. The PDR provided a chance for the team to get advice from others, including the Massachusetts Academy High School faculty, other neighboring teams, and professors from WPI who reviewed the computer models and physical prototypes to identify the strengths and weaknesses of the design.

Near the end of the construction process, a second formal review was held. This review, called the Critical Design Review, was conducted to receive feedback on the final design with the intent of completing any needed design changes prior to shipping the robot. Through the formal and informal review processes, the team received a continual flow of new ideas and careful examination to ensure no detail was overlooked. As demonstrated by their performance, Team 190's diligent application of fundamental design strategies proved to be a valuable method to design and construct a world-class robot.

▼ The secondary design goal was to lift tubes off of the floor and place them on all three levels of the rack. A lightweight and fast-acting mechanism, powered by a single piston for the tube manipulator and one motor for the elevator, proved to be robust and highly effective.

◄ The ability to independently lift both alliance partners at the conclusion of the match was the primary design goal for Team 190. They rationalized that many matches would be decided by the sixty points that were awarded for accomplishing this challenge.

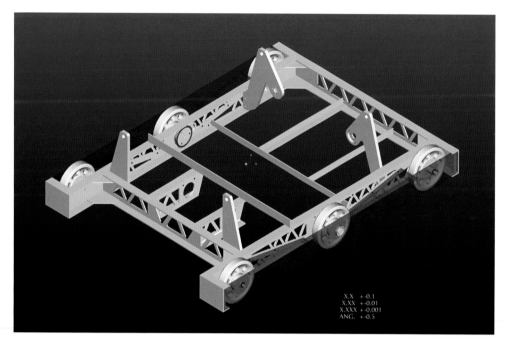

A six-wheel drive system, with the middle wheels offset below the outer wheels, was favored to achieve the high degree of maneuverability needed in the competition. Aluminum channel was selected as the chassis material to provide strength and rigidity.

X.X   +-0.1
X.XX   +-0.01
X.XXX +-0.001
ANG.   +-0.5

A POWERFUL PIVOTING DRIVE SYSTEM AND SOLID CHASSIS

Analysis and experience determined that the robot base would need to be stable, strong, and highly maneuverable. To maintain position and have an ability to be effective at defense, it was decided that the drive wheels needed a high degree of traction. A six-wheel drive system was designed to provide this traction yet also allow a high degree of maneuverability. By placing the center wheels of the drive system lower than the outer wheels, the robot would pivot on only four wheels at a time and be turning with a very short wheelbase.

A custom transmission was designed by the team to accommodate two motors on each side of the drive system. While the original drive system consisted of four-CIM motors (the most powerful motors in the FIRST Kit of Parts), weight limitations mandated that the weight of the drive system be minimized. The much lighter Fisher-Price motors replaced one of the CIM motors in each transmis-

sion. Though much lighter, the smaller motors produced 65 percent of the power of the CIM motors. Because these smaller motors operate at a higher rotating speed, an additional gear in the transmission was added to reduce the new motor's speed to the CIM motor speed. The alteration to the transmission was minor and only required one day to implement.

To withstand the inevitable on-field collisions, the robot chassis needed to be strong yet lightweight. Team 190 used aluminum channel as the chassis material to create a stiff frame that withstood shock loads and avoided warping. U-shaped aluminum channel was ideal due to its ability to resist bending and twisting. Individual channels were welded together to create a solid box frame for the chassis. To further prevent warping, gussets were riveted to each corner. Excess material in each channel was removed to reduce weight while preserving strength.

An auxiliary transmission was added to accommodate the higher speed and lighter weight Fisher-Price motor as a second drive motor. The single-stage reduction reduced the motor speed to match that of the larger and slightly more powerful CIM motor.

▲ The chassis consisted of
aluminum channel that was
welded together, with gussets
riveted to each corner to add
rigidity. Viewed here from
below, the cross frames offer a
foundation to mount electronics, pneumatics, and power
components.

A COMPACT MECHANISM TO SCORE TUBES

The tube scoring system consisted of three components: a claw to grab tubes, an arm on which the claw was mounted, and an elevator to raise tubes to each level. Bounded by the constraints to be lightweight and have a low volume, the design envelope on this system was extremely restricted. It was decided early in the design process that the claw would be pneumatically operated since the robot would be carrying a compressor and storage tanks to provide lifting power.

The original concept for the claw was a fixed plow to slide under the tube with a one-piece pneumatically operated gripper clamping down around the top of the tube. Testing revealed that this design tended to push tubes away and prevented positive control of the tubes. Insights from an engineering mentor who was working on a prosthetic arm and hand at Dean Kamen's DEKA Research and Development Company suggested that a four-bar linkage be used as a mechanism that could grasp the tube like a finger.

The design had multiple joints and could close around the tube's circumference, drawing the tube closer as it closed. Polyether-based viscoelastic polyurethane sheet covered the Lexan claw to further increase the ability to firmly hold tubes. The design was certainly innovative, and because it was powered from a single piston, the design was inherently robust as well. The fingerlike mechanism was capable of closing until it firmly grasped the tube—a feature that proved essential when handling tubes with varying levels of inflation.

The claw was mounted on the end of a miniature arm. Clever engineering, aided by CAD analysis of the four-bar mechanism, resulted in the ability to power both the claw and the arm with a single piston. The first part of the piston's stroke closed the claw and the remainder of the stroke moved the arm 45 degrees from its original horizontal position on the ground. Testing indicated that placing the tubes on the rack's spider legs at a 45-degree angle was the ideal orientation for fast and accurate scoring.

The design process for this combined grabbing and rotating system progressed from initial prototypes to CAD analysis that examined the piston placement, linkage lengths and positions, and moments created while lifting a tube. It was modeled and animated within the computer environment to demonstrate the viability of such an integrated system.

The miniature arm and claw were rigidly attached to the elevator, which extended to 72 inches (1.8 m). The elevator consisted of a three-tier nested aluminum channel where the outer channel was 1× 3 inches (2.5 × 7.5 cm) and the inner channel was 1 × 2 inches (2.5 × 5 cm). The third stage was 1-inch (2.5 cm) aluminum boxtube that served as a foundation for the miniature arm and claw.

▶ Modeled on the function of a human finger, the manipulator reached over and clamped down on each tube. The four-bar linkage provided contact with each tube around most of the tube's circumference.

▶ Extensive CAD analysis resulted in the ability to determine the size, orientation, and relation between components on the tube handling system. Computer-based animation and force analysis was used to design all linkages on the manipulator and the miniature arm to be powered by a single pneumatic cylinder.

▶ The aluminum channel was machined to reduce weight while preserving strength. The miniature arm and tube manipulator were mounted to the inner stage of the elevator to achieve over 8 feet (2.4 m) of vertical motion.

◀ The elevator lifting mechanism consisted of high-strength cord that was wrapped around a capstan. The cord ran inside the elevator and was fixed at the bottom of the innermost stage of the elevator. Running the capstan forward raised the elevator while running it in reverse pulled on a separate cord to lower the elevator.

▶ The ramp was designed to be a gradual slope that could be climbed by most teams. Once a robot reached the end of the slope the inner section of the articulated platform lifted to raise the alliance partner to a distance that earned a maximum score.

▲ To fit inside the competition's starting size envelope, the ramp hinged in two places and folded until needed. Over 2,000 holes were laser cut in the ramps' sheet aluminum and dimpled with a die to add structural strength to the lightweight structure.

A winch was designed and constructed to raise and lower the elevator. The winch had two lines that served as a pull-up cable and pull-down cable. Each line was 0.040 inch (1.0 mm) 300 lb.-test (1.3 kN) Spectra cord. The cord was run through a series of pulleys attached to each section of the elevator before being secured to the bottom of the innermost section. As the winch wound, the cable tightened and the elevator was raised. The weight of the elevator, claw and tube was only 9 lb. (4.0 kg).

THE MIGHTY RAMPS

The most important feature of Team 190's robot, as determined by their original performance goals, was the lifting ramps. It was clear that a ramp's effectiveness would be correlated with the slope of the ramp. To provide a continuous and gradual slope, Team 190 designed articulated ramps on each side of the robot. Once a robot climbed the ramp, the inner section of the ramp would be lifted using pneumatic cylinders to a height of 12 inches (30.5 cm). Like the claw, the ramps were simple in theory and elegant in their design.

Similar to the tube manipulator system, the ramp design was tightly constrained by weight and volume limits. Since the composite ramp was hinged, it could be carried on the robot in a folded condition. For most of the match the ramp remained in the close-packed orientation and was only unfolded at the end of the match when the robot was in the home zone.

Minimizing weight was critical for the ramps. A variety of materials were examined to construct the ramps including foam-aluminum composites and honeycombed aluminum. When examining the possibility of using sheet aluminum, the team discovered a process to stiffen the material

by dimpling it. Additional research identified the optimum dimple size and pattern orientation for their application. Over 2,000 holes were laser cut in the 5052-aluminum alloy $\frac{1}{16}$-inch (1.6 mm)-thick sheet. The holes were oriented with a hexagonal close-packed pattern to maximize the material's strength. Each of the ¾-inch (1.9 cm) holes was modified by forcing the aluminum into a predefined shape using pressure and a custom dimple die.

In the end, 20 lb. (9.1 kg) of sheet aluminum and rivets supported over 300 lb. (136 kg) of weight. Each ramp was lifted with three 1½-inch (3.8 cm) bore, 4-inch (10.2 cm) stroke pneumatic cylinders. Collectively, these pistons provided 318 lb. (1.4 kN) of lifting force. To prevent mounting robots from sliding off the ramps during climbing and lifting, antirollback devices were added to the ramps. A Lexan restraining device acted as a spring that folded down as a robot passed over it and popped back up once a robot passed. This simple feature proved to be extremely valuable and was credited with winning many matches.

DOING IT ALL, AND DOING IT ALL WELL

Team 190 benefited tremendously by establishing clear goals, utilizing a critical review process, and by a series of tests and prototypes to determine the best design features. Their years of experience, combined with research into material properties and manufacturing techniques, proved to be a winning combination that resulted in the highest performing robot in the 2007 FIRST Robotics Competition. Every aspect of their final design was closely scrutinized and tested, with each step leading the team along its overall goal of being innovatively competitive. Given Team 190's standing, they were able to do it all, and do it all well.

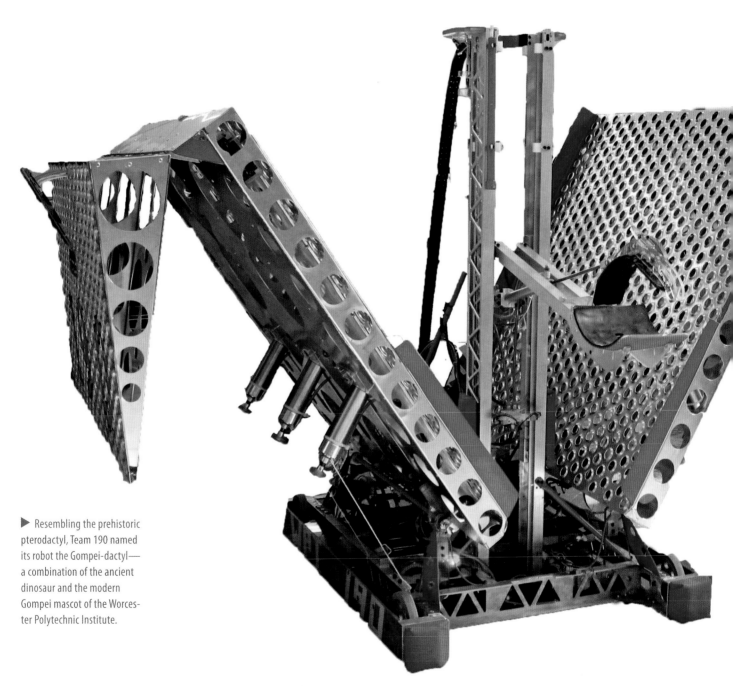

▶ Resembling the prehistoric pterodactyl, Team 190 named its robot the Gompei-dactyl— a combination of the ancient dinosaur and the modern Gompei mascot of the Worcester Polytechnic Institute.

Designing Quality into the Robot

Winning the Motorola Quality Award at the FIRST Robotics Championship was the season pinnacle for CyberBlue Team 234. Following their victory, the team initiated a post-season review to identify what went well, what needed improvement, and what processes needed to be discontinued.

This demonstrated commitment to continual improvement and high quality is a reflection of the corporate values of the team's sponsors: Rolls-Royce and GM subsidiary Allison Transmission. By using an industrial design methodology, Team Cyber Blue anticipated the competition needs, designed a robust machine, and constantly improved their design. This focus on quality, combined with an emphasis on simplicity and innovation, were the defining attributes of this award-winning team.

CYBER BLUE DESIGN—AN ONGOING PROCESS

A formal design methodology guided the team's activities to generate ideas and convert these plans into high-performing realities. The design process began with six steps that produced a design concept for the robot. The team first reviewed the rules to ensure that all team members understood the problem definition. Potential strategies were next developed in brainstorming sessions and a rubric identified the design features of each strategy.

The list of possible strategies and robot features was reviewed before deciding on a tube-scoring robot design. The robot would not have ramps, but would be able to climb a 30-degree incline. Concepts were presented and evaluated against the original design rubric to settle on a robot with a telescoping arm that would pivot on a turret. The design at this stage consisted of a simple chalk sketch, but the details behind that sketch were carefully examined, reviewed, and accepted by the team.

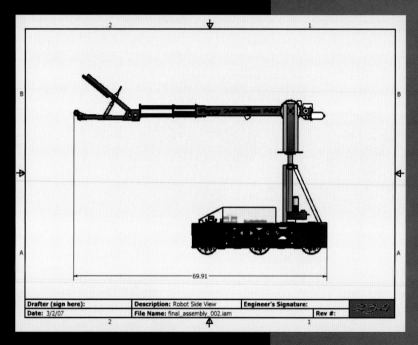

▲ An Autodesk Inventor model documents the developing design and is used to evaluate the robot's capability. In the presented configuration, the claw position is positioned at a height advantageous for scoring on the rack's middle row.

▶ A chalk drawing results from a careful analysis of the game requirements and the team's view of essential robot features. This sketch documents the team's initial thoughts that were refined and improved as the design process progressed.

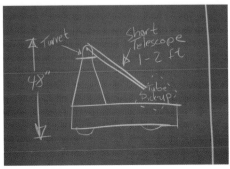

▲ The constructed robot closely matches the initial sketch and refined computer models. The simplicity of the subcomponents greatly enhances the machine's performance and durability during match play.

The design progressed from a chalkboard sketch to detailed plans using Autodesk Inventor software. Each of the designed component systems was then prototyped with any changes captured in the CAD drawings. A sequential process of design, prototype, test, and modify was constantly applied, with actual components manufactured and assembled at the completion of this phase.

Four weeks into the process, the students led a Critical Design Review meeting with senior managers and engineers from the sponsoring companies. At this review, the students presented their strategy and robot design, as well as the detailed drawings and manufactured parts for the major subsystems. This forced the students to step back and evaluate their work, and was a chance to receive valuable input from new perspectives. The build, test, and modify process continued as the finished robot was constructed and evaluated as a complete system.

Additional improvements were made at the competition where adjustments and modifications increased the team's competitiveness. The improvement process was continual and even extended beyond the competition. The final stage of the continuous improvement process, called Review and Improve, occurred at the conclusion of the season. A Root Cause/Corrective Action methodology was used to evaluate the year and create a lessons-learned document to improve the next season.

HIGH-TECHNOLOGY MANUFACTURING

In addition to using an industrial design process, Team 234 used industrial grade manufacturing techniques to design and build their robot. Accurate CAD drawings of the robot were created with a high degree of detail that ensured each system accommodated all other subsystems. For example, the CAD model of the base included the drive motors, bearings, transmissions, and wheels, with the frame designed to support these components.

Finite element analysis was conducted on the computer model of the frame to examine its capacity to distribute shock loads. Based on this analysis, vulnerable areas of the frame were reinforced to strengthen the frame. The computer model was also used to manufacture the frame with a computerized water-jet cutting machine.

▼ High levels of detail document the location and mounting requirements for the traction wheels, drive motors, two-speed transmissions, brakes, and inner drive wheels.

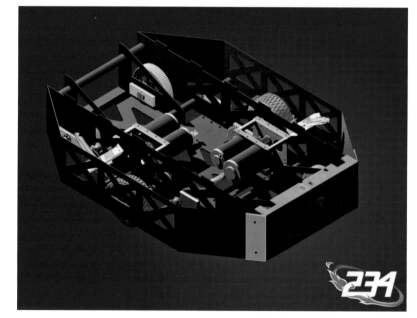

▶ For greater accuracy and higher performance, Team 234 uses water-jet cutting for the chassis plates. The material properties of the metal are preserved, as the water jet does not produce a weakened heat-affected zone on the cut pieces.

◀ A prototype chassis is designed and constructed during the off season to evaluate the durability and maneuverability of a new drive system. The prototype chassis is a valuable platform during the competition season because it served as a test platform for evaluating prototype mechanisms and control algorithms.

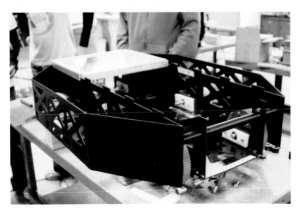

◀ Metal is used in all three chassis prototypes. The angled sides are designed to deflect and withstand side forces from any angle of attack.

THE ELEGANCE OF SIMPLICITY AND EFFECTIVENESS OF INNOVATION

Team 234 constantly pursued quality though simplicity and innovation. The claw is one example of the elegance achieved with simplicity as it consisted of two tubes operated by a single piston. With minimal moving parts and a fast cycle time, this simple mechanism was effective and reliable.

Innovations to the original claw design added a bottom hook to secure the tube and a conveyor belt to increase the claw's grip. Further innovations included adding two sets of wheels to the lower jaw. The wheels closest to the robot contacted the floor when the claw was driven into a tube. The second set of wheels was offset from the floor but in contact with the inner wheels. As the inner wheels rotated in one direction, the outer wheels rotated in the opposite direction to draw tubes into the claw. This innovation, which did not require an additional motor, was ingenious for its simplicity and effectiveness.

Other innovations improved the claw's performance. The claw was attached to the robot arm with a spring-loaded hinge. Because the claw was exposed in front of the robot, it was susceptible to damage from collisions with other robots. The hinge provided flexibility to divert the damage without impeding performance. To increase the strength of the claw, a second piston was added. A limit switch was added to the mouth of the claw to automatically close the claw when a tube was within its grasp.

Pneumatic power activated a set of brakes on the center drive wheels. Rectangular brake pads were mounted on the end of small pistons positioned over the center drive wheels. When activated, the pads pressed against the wheels and locked the robot in position. These brakes kept the robot in a fixed position when scoring tubes and anchored the robot in place after climbing on top of an alliance robot.

▲ The prototype claw is mounted on the prototype chassis as a test platform to investigate scoring methods. This operating platform is also used by the programming team to create and test autonomous-mode computer code.

◀ A single piston opens and closes the prototype claw. The bottom tube remained fixed and parallel to the ground, with the upper tube pulling down on a tube to secure it in the claw.

▼ A second piston is added to increase the claw's holding power. The small piece of vertical hose in front of the piston is a trigger that is activated by a tube and automatically closes the claw.

◀ An accurate CAD model includes the passively powered wheels to draw a tube into the claw and the hinge where the claw mounts. The hinge adds flexibility to the claw and allows the claw to rotate and avoid damage if impacted by other robots.

▶ A small piston mounted on the side of the traction wheel serves as a brake pad. To hold the robot in position, air is supplied to the piston and the brake is tightly held against the wheel.

QUALITY DESIGN AND INNOVATION

Team 234's journey from a chalkboard sketch to an exact CAD model and multiple prototypes—used as templates for constructing an elegant machine—was not an unguided adventure. Rather, Team Cyber-Blue followed a very specific plan of action that included frequent review and enhancements. Simple mechanisms with a minimum number of moving parts increased the system reliability and innovative ideas perfected the robot's performance. Team 234 diligently applied these principles to produce a high-quality robot that rose above the competition to win the 2007 FIRST Championship Motorola Quality Award. One can only guess what improvements they are already planning for next year.

▶ The robot's reach is extended using pneumatically powered slides. This capability enables Team 234 to reach over a defending robot and score, even if the defender was positioned between Team 234 and the rack.

Quality Sparks Strong Performance

▼ Sparky 8 is an omnidirectional robot that uses suction cups and a three-stage elevator to place tubes on the rack. Earlier plans to include a ramp on the robot were sacrificed to keep the robot within weight limits and to emphasize simplicity over expanded functionality.

Simplicity and quality have always dictated Team 384's design and build processes. These principles guided every design decision, starting with the team's preseason planning sessions and continuing to the final match of the year. Rather than waiting until January when the 2007 game was announced, Team 384 started their design process in September when they investigated competition aspects that were not completely dependent on the game. Though the FIRST competition rules require the robot be constructed after the game is unveiled, there is no restriction on when teams can think, learn, and plan.

▼ A motor directly powers each of the four drive wheels to create a highly maneuverable platform. Dual side panels support each end of the wheel drive shafts and create a strong frame capable of withstanding intense robot interaction.

During the autumn, Team 384 investigated and constructed a prototype base to explore the use of an innovative drive system. After the game was announced and game-specific manipulators were designed, they were added to this base to investigate their effectiveness. Throughout the process, the team was guided by a desire to conserve weight and achieve superb durability and mobility.

Team 384's robot, named Sparky 8 in recognition of the team's eight-year history with FIRST, was designed to be a tube scorer. Tube control was obtained with suction cups powered by a venturi-based vacuum system. A chain driven, three-stage elevator lifted the tubes to score on all levels of the rack. A team-designed transmission was directly connected to a unique set of wheels that allowed the robot to move in any direction on the field without changing its orientation. These simple systems were assembled with a high level of quality to produce a dependable and robust robot.

INSPIRATION AND TESTING IDEAS

Team 384 drew on many different sources of inspiration, ranging from their sponsor's industrial

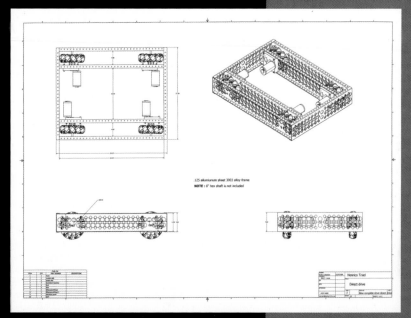

background, other teams, and their own experience building competitive robots. The drive system used commercially available Mecanum wheels. These wheels allowed simultaneous rotation on two axles and enabled the robot to move in any direction. Team members attending the 2006 FIRST Championship witnessed the effectiveness of these wheels and convinced others of their benefits for Team 384's 2007 robot.

The team's own experience with chain-driven transmissions prompted the use of a direct-drive transmission. A direct-drive transmission had fewer moving parts and was more compact—both important design attributes.

One of Team 384's sponsors was Flexicell, a company involved in packaging equipment and pneumatic systems. Engineering mentors from Flexicell recommended industrial-quality suction cups manufactured by their company as the means to lift tubes. Cranes and forklifts were also industry-based applications that were models for the robot. After studying these systems, a forklift style lift was chosen as the best method to incorporate on the robot.

The robot base and drive system, developed in the autumn, served as a test platform for investigating subsystems on the robot. Working prototypes of each of the various systems were evaluated independently as an integrated component to identify the best combination of systems. The prototypes often started as static wood models before being powered with motors and pneumatic actuators. Working in wood was a fast method to test ideas and confirm space needs for each part.

In addition to maximizing performance, the prototype process was also valuable as a means to minimize weight. The tests illustrated parts that weren't necessary for the system's function and could be removed without affecting performance. As the parts moved from their prototype phase to their final form, they were lightened to save weight by removing material that did not add strength. As one example, the frame was reduced from 17 lb. (7.7 kg) to 10 lb. (4.5 kg) by punching holes in the aluminum side panels. The holes reduced the weight but did not compromise the frame's strength or durability.

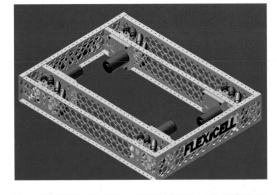

▶ The weight of the aluminum frame is significantly reduced by punching holes throughout the side panels. The frame rails are spaced to accommodate the Mecanum wheels.

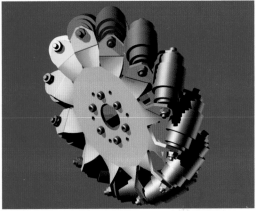

▶ Commercially available Mecanum wheels provide a high degree of reliability and create a highly maneuverable robot.

▼ Each wheel is carefully examined and adjusted to optimize performance. The tension on each roller is modified to allow the rollers to freely spin while keeping the rollers firmly in place.

DESIGNING A DIRECT-DRIVE TRANSMISSION

The drive system consisted of two important components: the transmission and the Mecanum wheels. Mecanum wheels were favored for their increased maneuverability. The drive system works by changing the direction of rotation for each wheel. Rollers on each wheel's perimeter are free to rotate as each wheel spins. The combined wheel and roller actions allow the robot to move sideways and diagonally, as well as forward and backward.

A direct-drive transmission was designed as a compact, maintenance-free power source for the Mecanum wheels. Because each wheel was independently powered, any weight or space savings that could be achieved with an efficient transmission design had a multiplied impact.

The direct-drive transmission was favored over a transmission that would be used with a chain and sprockets for two reasons: the direct drive had the potential to use fewer parts and would therefore be lighter, and the direct drive would require less maintenance during competitions. The team's past experience with chain-and-sprocket drive systems was less than perfect as prior designs required continual adjustment to keep the chain tight, and often included reattaching chains that had fallen off during matches.

The direct-drive transmission used four gears to reduce the drive motor speed in two stages. The transmission's four gears were designed and fabri-cated by team members. Accurate CAD models of each gear were entered as input data in multiple computer-controlled machines to cut each gear. Similarly, other components such as the transmission end plates and wheel hubs were designed and fabricated by team members. Assembly plans for the transmission were also created in CAD to document the assembly of each transmission. Coupled with the drive motor, the computer models were integrated into the computer model of the frame to ensure the transmissions were easy to mount and fit within the assigned space.

Team 384 was able to realize other benefits of designing their own transmission, in addition to minimizing weight and space. Because the team manufactured their own gears, they were able to obtain the exact speed reduction desired for the drive system. In addition, team members were intimately familiar with all components of the gearbox and could easily troubleshoot and maintain the transmission.

By working on the drive system during the autumn, the team was able to perfect its design and have a working robot platform to investigate game play during the competition season. The team used its knowledge to perfect and construct the final version of the drive system in a very short period during the January build window as specified in the FIRST competition rules.

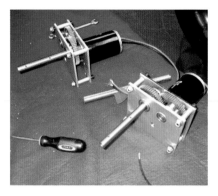

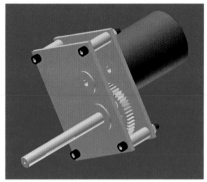

◀ The gearboxes increase the delivered torque by stepping down the speed of the drive motors with a two-stage gear reduction. Long output shafts allow the wheels to be directly mounted to the transmissions to create a compact drive system.

▶ An exploded view of the custom gearboxes is created using Autodesk Inventor. The diagram serves as an instruction manual for student team members to assemble each gearbox.

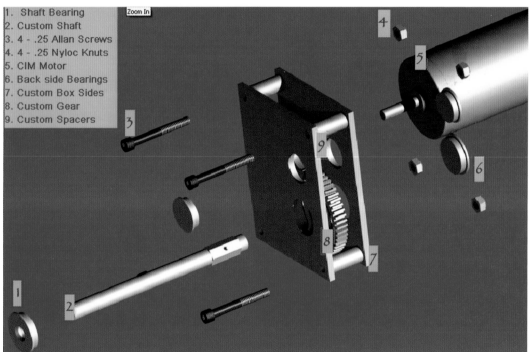

1. Shaft Bearing
2. Custom Shaft
3. 4 - .25 Allan Screws
4. 4 - .25 Nyloc Knuts
5. CIM Motor
6. Back side Bearings
7. Custom Box Sides
8. Custom Gear
9. Custom Spacers

Zoom In

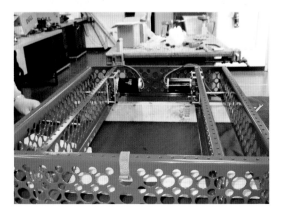

◀ The frame is constructed with aluminum channel that is lightened and powder coated. The transmission is mounted directly to the frame—a measure that conserved space and avoided additional mounting hardware.

QUALITY FEATURES ABOUND

Team 384's focus on simplicity was the motivation behind many design decisions. Commercial Mecanum wheels were purchased and modified to meet the team's specific needs. Custom wheel hubs were manufactured to mate the wheels to the transmission driveshafts. Commercial aluminum channel was selected as the frame material and welded together to create the base. In both of these cases, off-the-shelf components were selected and modified—a process that saved time and ensured performance. The base and wheel hubs were powder coated to increase the robot's aesthetic appeal.

Other design details helped increase reliability and improve performance. Pins were used instead of shaft collars to keep the wheels in place on the driveshafts, and the custom hubs were designed with long keyways to securely transmit the motor power to the wheels. The electrical system was constructed on a removable panel that, once completed, was installed in the base of the frame. Because the wheels, transmissions, electrical panel, and tube manipulator were modular components, each could be separately constructed and tested before being integrated into a fully functioning robot.

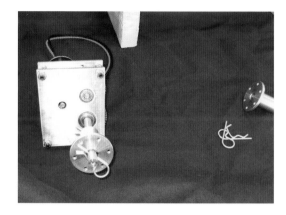

▶ Internal wheel hubs are designed and fabricated to mate the Mecanum wheels with the transmission driveshafts. The hubs are keyed to the shaft and held in place with pins that could be easily removed if the wheels or transmissions need repair.

▶ The frame protects the Mecanum wheels from damage, with additional protection provided by the robot's bumpers. Holes for the retaining pins in the driveshaft are carefully placed to keep the wheels in place and prevent lateral slipping.

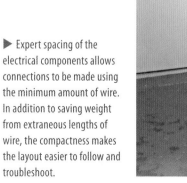

▶ Expert spacing of the electrical components allows connections to be made using the minimum amount of wire. In addition to saving weight from extraneous lengths of wire, the compactness makes the layout easier to follow and troubleshoot.

The completed electronics board is mounted directly to the frame using pre-existing holes in the frame's rails. Each drive motor, transmission, and wheel is accessible from the bottom of the robot—a design detail that eases maintenance procedures for these critical components.

Simplicity and quality guided Team 384's design and construction practices to produce a highly maneuverable and reliable robot named Sparky 8.

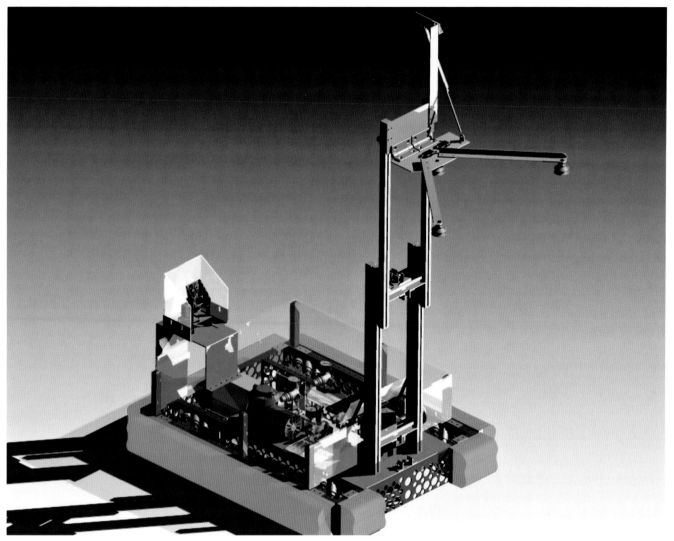

Transformer: Ramp-bot in Disguise

To the uninformed observer, the starting configuration of Team 1114's robot might have appeared to be a congested nest of odd parts haphazardly connected together. Many of the components were stored at what were seemingly odd angles with a billboardlike structure leaning out, a box on the bottom skewed up, and a set of rails titled back. A vertical tower of intertwined components was the only section mounted at a normal angle.

But as the match began, the collection of parts unfolded and transformed the tightly packed components into an ever-expanding system of integrated robot functions. An arm and claw emerged from the vertical tower, having been packed inside the tower's outer supports. An elevator mounted in the center of the tower lifted the arm to provide the machine with additional reach.

Most impressive was the expansion of the odd-angled components, with one structure falling back, another rotating forward, and the sides falling outward to build a ramp. During this time, the arm repacked itself as a vertical tower and bowed down at the front of the robot. The final transformation occurred at the end of the match when the angled ramp morphed into a horizontal platform to lift its alliance partners off the playing field. In this configuration, the robot was very different from its original compacted configuration—a transition that was possible only with thoughtful planning, purposeful integration, and imagination.

▲ In its starting configuration, Team 1114's robot is a tightly packed system of systems. With few motors located outside the base, the robot is very stable.

ROBUST DESIGNS

Team 1114's strategy was to score as many ringers as possible and lift robots for bonus points. A strong scorer would attract defenders, making it imperative that the robot be designed with robustness as a priority. A successful robot needed to be designed to withstand interactions with its opponents and avoid mechanical failures.

To achieve robustness, care was taken to ensure that all mechanisms were as simple as possible. The emphasis on simplicity promoted the creation of systems with the fewest number of parts. Also, a simple design would be much easier to operate because the number of control options would be minimized. The expected game play prompted a careful analysis of robot components that would undergo repeated stress. For example, because the arm joints were suspect to damage yet needed to always function, they were designed with strength and endurance in mind. Similarly, the drivetrain was designed to be stiff and impact resistant to withstand the expected constant interaction with other robots.

The size and weight constraints were other factors critical to the design process. Staying under the 120-lb. (54-kg) weight limit required that each component be designed to minimize its weight. The size constraint dictated that each mechanism be integrated with every other component.

With an equal emphasis on scoring tubes and lifting other robots, it was essential that the mechanisms for these systems be compatible with each other. This requirement resulted in a very low drive base, a multisection ramp stored at the rear of the robot, and a compact arm that was capable of moving out of the ramp's way. The integration of each component with the others was one of the most challenging aspects of the design phase. The complex amalgamation was only possible because the individual mechanisms were designed from the very beginning to meet form and function requirements.

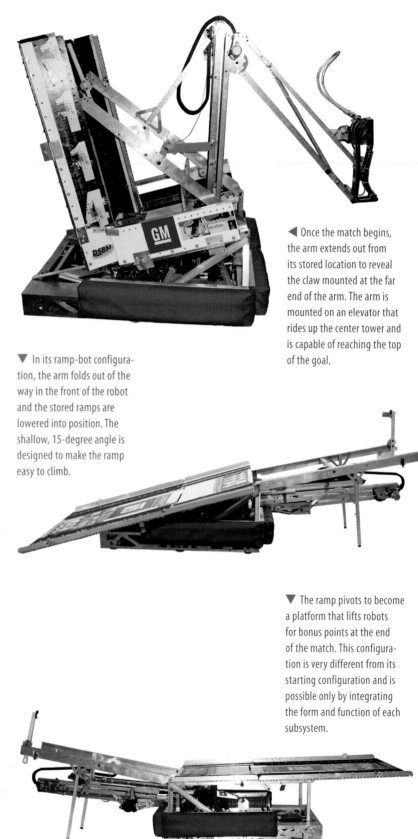

◀ Once the match begins, the arm extends out from its stored location to reveal the claw mounted at the far end of the arm. The arm is mounted on an elevator that rides up the center tower and is capable of reaching the top of the goal.

▼ In its ramp-bot configuration, the arm folds out of the way in the front of the robot and the stored ramps are lowered into position. The shallow, 15-degree angle is designed to make the ramp easy to climb.

▼ The ramp pivots to become a platform that lifts robots for bonus points at the end of the match. This configuration is very different from its starting configuration and is possible only by integrating the form and function of each subsystem.

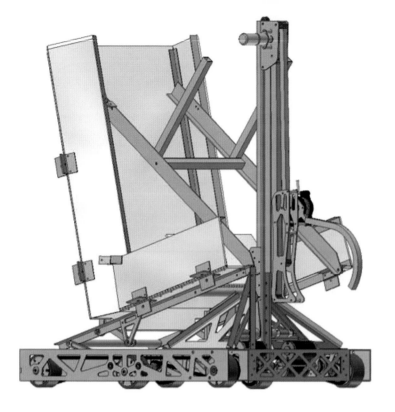

Kinematic analysis is conducted using CAD to determine the optimum gear sizes of the arm's shoulder joint. The small area available for the gears also has to support the motor, gear shafts, and pivots for the linkages, and connect all of these components to an elevating rail.

▲ Designed as a four-bar linkage, the arm's narrow profile allows the system to fit inside the robot's starting volume. The low height of the base is sized to support the ramp as it unfolds from its stored position to its shallow climbing angle.

THREE STATES OF THE FOUR-BAR LINKAGE

The arm needed to exist in three separate states: as a narrow, compact column to start; as a multiple-length appendage for scoring; and as a folded-down mechanism to accommodate the ramp. The beginning and end states complemented each other as a narrow tower could conceivably be folded down if its base was attached to a mechanism that allowed rotation. That requirement did demand that the rotating mechanism be strong enough to support the arm during the bulk of the match when the arm was extended.

The inspiration for using a four-bar linkage for the arm originated from a team mentor who applied his experience from an engineering mechanisms course to design the linkage and achieve the desired range of motion. The four-bar linkage enabled the arm to be tightly packed at the beginning and end of the match. In its nested position, the shorter end linkages allowed the longer linkages to rest on top of each other in a minimum amount of space.

The four-bar linkage extended using a single motor mounted to the inner linkage. A larger gear attached to one of the longer linkages provided the needed torque to rotate the arm. The gear sizes were determined based on the torque needs and the size constraints of the stored system. A wide

connecting link provided ample room for mounting the motor and two shafts that supported the gears. This connecting link was attached to a bar that rode up and down the tower. When fully raised, the arm reached the top level of the goal and when retracted, the manipulator picked up tubes off the floor. The bar linkage served a dual purpose of connecting the links and providing a foundation for the tube pincher.

Weight was an important consideration in the arm. Structural members were made from aluminum tube with a wall thickness of $\frac{1}{16}$" (1.6 mm). The center tower was strengthened with tightly fitted spruce inserts that were hammered into the aluminum tube. The resulting composite structure had significantly less bending than the same member without the spruce reinforcement. Nylon blocks were inserted into the ends of the four-bar linkage to provide strength at the arm joints. The nylon distributed the load along the length of each pivot point.

The four-bar linkage arm prompted a series of design simplifications. An original design called for an ability to rotate the tubes, but this version was simplified by modifying the lengths of the four-bar linkage to change the angle of the tube as the angle of the arm changed. The desired tube rotation function was achieved and as a result, a simpler gripper mechanism could be used. The arrangement of the individual arm pieces caused the widest separation between the arms to occur when the arm was at its maximum reach to provide the greatest stability.

▶ A complex system consisting of an elevator, arm, and claw unfolds from its packed position at the front of the robot. The four-bar linkage is designed to rotate the tube as the arm rises, making it easier to place the tube on the goal.

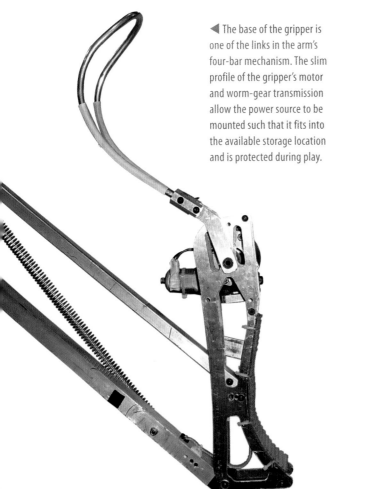

◀ The base of the gripper is one of the links in the arm's four-bar mechanism. The slim profile of the gripper's motor and worm-gear transmission allow the power source to be mounted such that it fits into the available storage location and is protected during play.

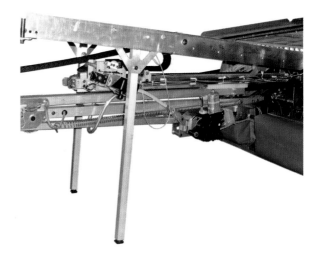

▲ The complex arm and gripper are designed to rotate into a horizontal position at the end of the match to transform the tube-scoring robot into a robot-lifting machine. In the prone position, a pair of rails protect the arm from damage if an alliance robot falls off the ramp.

UNFOLDING A RAMP AND LIFTING A PLATFORM

The ramp was designed to lift two robots 12" (30.5 cm) off the floor. Its width was sized to allow room for the widest robots to climb on it. The width also was a benefit for narrow robots because it required less precise driving. The shallowness of the ramp eased climbing and was not so steep that robots would roll off once they climbed into position. The lift mechanism that converted the ramp into a platform was designed to be fast so that teams could wait until the last moments of the game before climbing.

The weight and strength of the ramp were major considerations. Various ramp materials, such as carbon fiber and composite panels, were evaluated before selecting aluminum honeycomb. The honeycomb was chosen because it best satisfied the strength and weight criteria and was readily available at an affordable price.

A series of steps unfolded the ramp and raised it to become a platform. The largest section of the ramp, which included the edges that were folded in and a guide rail to align climbing robots, was mounted to the base with a four-bar linkage. As the linkage expanded in length, the ramp fell from its vertical orientation. When the far end of the ramp was in contact with the floor, this section of the ramp was at a 15-degree angle and was in line with the smaller ramp section on the robot's base.

At the front of the robot, the arm folded down and a set of rails extended out over the arm. Once in this configuration, alliance robots were allowed to climb the ramp. The lifting mechanism was based on a motorcycle kickstand design that allows the bike to be pushed forward to lift the back wheel off the ground. A winch was used on the robot to pull the ramp supports into a vertical position where they locked in place. The overall ramp design was a bit more complex than that of many other teams, but its shallow angle made it possible for a greater number of teams to climb it.

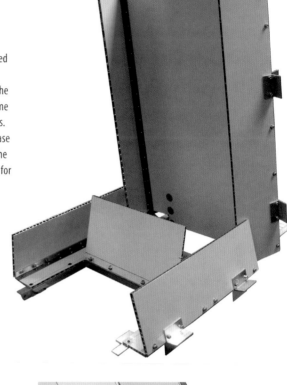

▶ The origami-configured ramp is folded in numerous dimensions to pack the platform inside the volume allowed in the FIRST rules. The sides unfold to increase the ramp's width while the vertical sections fold out for maximum length.

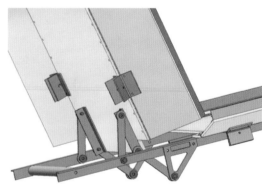

▶ Four-bar linkages are used in the ramp to guide the pieces from a stored orientation into a scoring configuration. The system is modeled in CAD to determine the correct linkage lengths to achieve the needed range of motion.

▶ The ramp is transformed into a platform using a winch that is attached to supports connected to the bottom of the ramp. In the stored position, the supports rest inside the robot base. Powered by the winch, the supports ride on rails in the base and rotate to a vertical position to lift the ramp.

TRIPS-ONE-FOUR: MORE THAN MEETS THE EYES

Combining such complex mechanisms in a small area required clear communication as each part was designed and constructed. An online CAD repository was created to record and share the arm, ramp, and drive subsystem designs. The repository was continually updated and changes were communicated to the designers of each subsystem. Logged comments detailed each change and the reasons for the alteration. Because each system was linked to every other subsystem, communicating the changes among all team members was an important factor in the team's success.

The configuration of Team 1114's robot at the end of the match was much different from its configuration during that match, and even more different than the configuration at the start of the match. The robot was designed to evolve into arrangements unique to each part of the game. As a transformer robot, Team 1114 had many hidden features that were revealed during the game. The unveiled variety of configurations was certainly more than meets the eyes.

▲ The base's low profile requires an efficient layout for the electronics. The rails for the ramp supports are placed on the outside edges of the base.

▶ In its final configuration for lifting alliance robots, Team 1114's robot looks very different from its original configuration when all components fit inside a 3' × 4' × 5' (0.9 × 1.2 × 1.5 m) starting volume.

Robot-Works: The Invisible Robot

▼ With Lexan as the principle building material, Team 1714 created a robot that bared it all. The material provided structural strength and transparency that allowed unobstructed views of the robot's internal components.

Team 1714 was motivated by a fun goal: to create a really cool and unique robot that could sustain impacts during competition without breaking. The team expected a significant amount of smashing and bashing and wanted a robot that minimized the number of repairs needed between matches. As an example of the desired resilience, the robot would be designed to sustain a full-speed collision with another robot without damage.

Lexan was chosen to construct the robot because of its strength and the clean aesthetics of a clear machine. This material was favored to help the robot stand out among the field of primarily metal machines. There was a strong desire to create a clear robot that you could look through to see all internal components. As a transparent robot, Team 1714 would be noticed for its uniqueness. The team hoped that the transparency, coupled with a strong performance, would gain the respect of other teams at the competition.

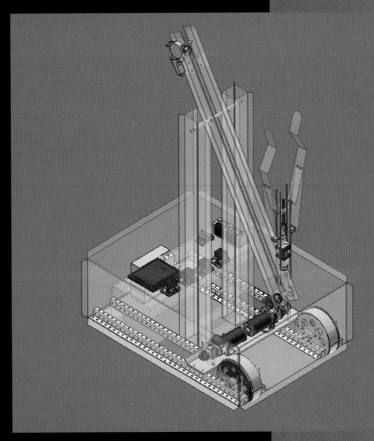

▶ Two wheels powered the robot using a three-speed transmission. The robot's motors, gears, shafts, and wires stand out amidst the many Lexan structural pieces.

POLYCARBONATE CONSTRUCTION TECHNIQUES

Creativity, experimentation, and testing helped Team 1714 develop construction techniques for designing a polycarbonate robot. As a plastic, this material is ideal for robotics because it is strong but not rigid. The flexibility allowed for parts, such as the side of the chassis, to deform upon impact and return to their original shape. The flexing distributed the applied load over a larger area.

Pieces were joined by screwing edges together that were bent up to a 45-degree angle. The edges allowed the material to slightly deform when loaded. In addition, the bent edges increased the stiffness of the Lexan plates and allowed thinner material to be used. A 6'' (15.2 cm) polycarbonate cube was constructed to examine possible manufacturing techniques. The prototype piece was also evaluated for durability.

The tests indicated that twisting was a problem if any side of the cube was removed. Additional tests showed that the cube retained the majority of

its strength even if 50 percent of the material was removed as a single hole in each cube face. This test confirmed the value of the 45-degree edges as they acted as beams to transmit the loads along the sides of the cube. The strength of a larger polycarbonate box measuring 25'' × 35'' × 10'' (63.5 × 88.9 × 25.4 cm) was tested and found acceptable. This box became the robot chassis. The same fabrication methods were applied to build elongated four-sided polycarbonate tubes that became beams for the robot's arm.

The flexibility that aided the Lexan's ability to accept high loads negatively affected the material's capability as a base for the drive motors. The length of the chassis allowed bending. To address this problem, aluminum rails from the Kit of Parts were fixed to the bottom of the chassis to stiffen the base such that the drivetrain could be installed. With their predrilled holes, the rails were a convenient material on which to mount the drive wheels, front castor wheels, and the drive motors.

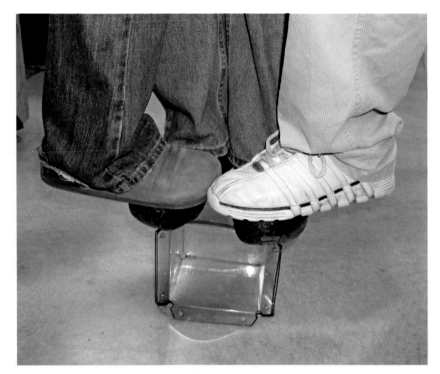

► Human load tests confirm that the transparent plastic has sufficient strength to serve as a robot construction material. The cube faces are attached to each other with small screws.

► The angled-edge construction method is used to build the structural components of the robot, including the base and towers that support the robot arm. The protective coating that the Lexan is shipped in is kept on the material to prevent scratches while the robot is built.

▲ Aluminum rails along the bottom of the chassis add stiffness and serve as mounts for the robot's drive system. Because the edges are screwed together, individual sides can be removed easily for access to internal components.

PAPER TO PLASTIC: THE ARM'S EVOLUTION

Team 1714's robot featured a double-jointed arm supported by a pair of columns rising from the center of the robot chassis. The arm design originated as a hand-drawn sketch that was refined as a CAD drawing. Based on the CAD model, the arm parts were fabricated with Lexan and mounted together. The final stage included mounting all the arm's power systems and the needed wiring.

The center columns were mounted to the aluminum rails in the chassis and connected to each other with polycarbonate bars at the columns' base. Additional connections to a raised electronics tray and the upper side of the chassis added stiffness to the columns. The towers provided the mount and pivot for the second section of the arm, which in turn served as the pivot for the third section of the arm that included the manipulator.

When the constructed arm was first tested, the motor could not produce enough torque to rotate the arm. To reduce the needed torque, a longer arm was constructed. Because of the relative ease of working with Lexan, the new arm was quickly constructed. The longer arm reduced the distance between the pivot and the section's end while preserving the arm's needed range of motion.

The end of the longer arm served as a method to effectively add counterweight to the arm. The motor powering the last section of the arm was located behind the pivot point in the longer arm. A plastic chain connected the motor and the upper-arm joint. Latex surgical tubing was used as a spring between the arm's end and the towers to pull the arm down and reduce the motor force needed to rotate the arm. Collectively, these changes reduced the load on the arm and allowed the arm to be rotated with a single motor mounted at the pivot point.

CLEAR CLAW CONCEPT

The concept for the claw originated as a sketch that illustrated how a threaded screw could be used to open and close a claw. A motor was housed in the claw's base. Fixed length members were attached to the base near the claw and connected to the claw's arms. Hinges were used as connectors to allow the claw to open and close when the threaded shaft rotated. Restraining barriers were added to the claw to keep any loosely held tubes from sliding down the arm.

This internal grasping design was favored over an external mechanism to grab the tubes for multiple reasons. The closed claw had a slim profile that provided a large margin when inserting the claw in a tube's center. The design also increased the tube's center diameter as the force to hold the tube increased—a feature that made scoring easier with securely held tubes. Because only thin plates held tubes along the inner diameter, the device did not interfere with scoring.

The flexibility of Lexan also benefited the claw's operation. The greatest benefit of using Lexan was discovered while scoring. The friction between the claw and the tube was sufficient to hold the tubes, yet allowed tubes to be pulled out from the claw during scoring. This meant that the claw did not have to be opened to release the tube—a design feature that speeded scoring.

Also, the Lexan easily withstood impact from other robots and simply flexed to deflect contact. Though the Lexan claw looked elegant, its transparency made it hard to use when trying to collect tubes at the far end of the field. To correct this problem, the outline of the competition claw was painted to make it easier to determine if the claw was inserted into a tube's center.

▶ Sporting the white protective film, the individual pieces of the arm are mated together to examine their compatibility. Still as a prototype, the last section of the arm is wider and bulkier than its final form.

TRANSPARENCY TURNS HEADS

Team 1714's robot—composed of a polycarbonate chassis; a center-mounted, double-pivot arm; and a three-speed transmission—was unique for its construction material and tube-holding mechanism. In addition to its elegant appearance, the robot was well designed and professionally constructed. The multiple drive speeds provided high torque that eased maneuvering and high speed to quickly transverse the field.

The most unique attribute of the robot was its transparency—a feature that constantly drew attention and interest. By using Lexan as the primary construction material, the team designed a robot that looked very different from any other competitor. The transparency resulted in a design that was hard to see, but increased the desire to be seen.

▶ The completed arm is much longer than the original version. The section behind the pivot point acts as a counterweight to reduce the torque needed to rotate the arm. Latex surgical tubing acts as a spring to further reduce the arm's load.

▼ The claw's fingers are shaped to squeeze the tube from the inside. As the grip tightened, the internal opening of the tubes increased—a design feature that increased the tube's diameter and aided

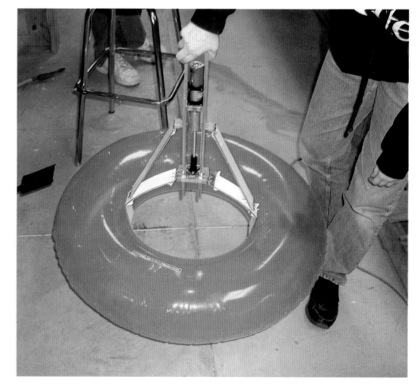

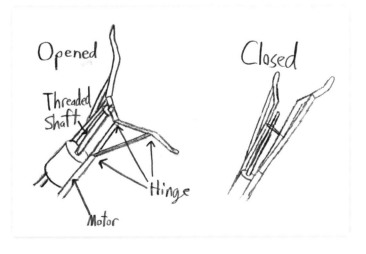

The claw's frame is mounted on the arm. Barriers are added to the claw to prevent captured tubes from falling down the arm. An internal sprocket at the claw's pivot is connected with plastic chain to a drive motor housed in the opposite end of the arm.

The first sketch of the claw proves to be an accurate model of the system that is ultimately constructed. The motor rotates a threaded shaft through a fitting mounted on the claw. As the screw rotates, the claw unfolds and expands as it pushes on the tube's inner edge.

Opened

Threaded Shaft

Motor

Hinge

Closed

Rockwell Automation Innovation in Control Award

Robot control is an important aspect of performance. The Rockwell Automation Innovation in Control Award celebrates an innovative control system, or an application of control components, that provide adequate machine functions. Robot functions must be monitored with sensors and measured using the robot control system. Control algorithms use the measurements to determine robot actions and correct for any discrepancies between the commanded signal and the resulting action. Recognition is awarded for performance during the game's autonomous period (where robots operate without human drivers) and during the teleoperated period of the game when operators control robot functions.

The autonomous portion of the 2007 competition challenged teams to design and implement control systems that located the rack, maneuvered robots to the rack, and scored tubes. The scoring rack was rotated before each match to prompt teams to use feedback control. Target lights above half of the scoring columns indicated the position of the rack and the best teams made use of this signal to find the goal and score.

The Killer App for the Killer Bees

With its dual-jointed arm, swivel wrist, and double-articulated claw, Team 33's robot required simultaneous control of its motors and pneumatic actuators. For manipulating tubes, three motors needed to be independently operated, with each motor affecting the operation of the others. Anticipating that the complex system would be very difficult for human operators to master, a control system was designed to direct the motion of each component. Pushing a single button commanded precise maneuvers to be executed, with feedback control guiding the action of each motor.

For Team 33, the Killer Bees, feedback control was the killer application for the 2007 FIRST season. The inclusion of controls in the robot design process eliminated the need for weeks of driver training: weeks that were hard to come by given the aggressive FIRST Robotics build window schedule. The robot, named Buzz12 to acknowledge the team's twelve years of FIRST involvement, was designed from the onset with effective control as a primary design goal. In addition to the programmed control, components were designed for positive physical control of the tubes to enable quick pick-up and scoring.

▶ The position of Team 33's elbow, shoulder, and wrist joints are monitored and controlled to accurately position the claw. Stability is maintained by measuring the robot's vertical angle and automatically directing drive functions to keep the robot upright.

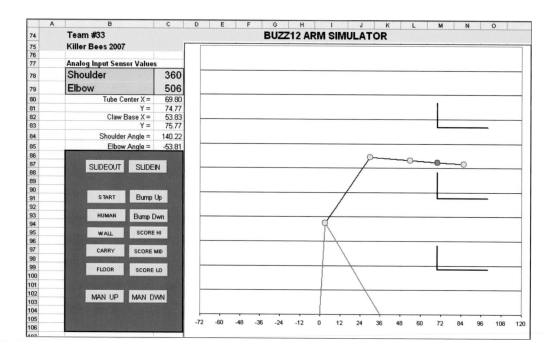

The following table is part of the image:

	A	B	C	D	E	F	G	H	I	J	K	L	M	N	O
74		Team #33						BUZZ12 ARM SIMULATOR						
75		Killer Bees 2007													
76															
77		**Analog Input Sensor Values**													
78		Shoulder	360												
79		Elbow	506												
80		Tube Center X =	69.80												
81		Y =	74.77												
82		Claw Base X =	53.83												
83		Y =	75.77												
84		Shoulder Angle =	140.22												
85		Elbow Angle =	-53.81												

Buttons: SLIDEOUT, SLIDEIN, START, Bump Up, HUMAN, Bump Dwn, WALL, SCORE HI, CARRY, SCORE MID, FLOOR, SCORE LO, MAN UP, MAN DWN

SOFTWARE APPLICATIONS SPEED THE DESIGN PROCESS

The Killer Bees established a goal to complete all software creation before the robot was built. A variety of software programs were used to analyze the robot, its components, and the resulting motion, as well as to develop the control processes that directed fluid motion of the robot arm.

A custom Visual Basic application in Microsoft Excel was written to simulate the arm position. This application developed the control geometry and algorithms to maneuver the arm along ten paths for scoring, grabbing a tube, and carrying a tube. The program prescribed the sensor values at the arm's shoulder and robot for each maneuver, and these variables were to write the computer code for feedback control. Once the robot was built, the sensors were calibrated and the feedback control gains were tuned for optimal performance. A custom National Instruments LabVIEW utility was written during this tuning phase.

To develop the robot's autonomous functions, another Visual Basic application was created to create the robot's path of motion based on a prescribed starting position and a desired end location at the rack. The software specified the coordinates for the robot to follow and predicted drive motor settings to move the robot along this path.

All parts of the robot were designed in Autodesk Inventor prior to fabrication. A complete set of design layouts and detailed engineering drawings were produced for the entire machine and all its components. The drawings indicated each component and documented how the parts combined to create individual mechanisms.

An Autodesk Inventor model confirmed that Team 33's robot met the maximum size constraint of 72" (1.83 m). The three-dimensional model was placed inside a sizing box. The software generated the shadow image of the robot and reaching arm to document adherence to design standards. The team's proficiency with this wide variety of software programs greatly enhanced their ability to develop the robot's commanded algorithms before the robot was built.

▲ A Virtual Basic simulation tool written using Microsoft Excel determines the resulting angles at each joint when the arm is moved to specific scoring positions. The tabulated data serves as input to direct motion on the actual robot.

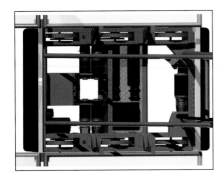

▲ Autodesk Inventor models determine the location of every component in the robot. Detailed plans sped development and preempted difficulties with integrating separate systems.

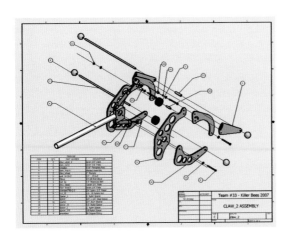

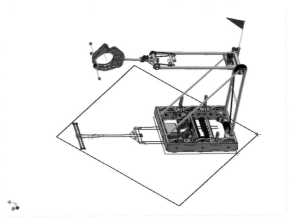

▲ Detailed drawings document how parts fit together to build complex mechanisms. The claw consists of a cam-driven upper finger and a lower thumb, each powered by a piston to tightly close around a tube.

▲ Each robot is restricted to a maximum reach of 72″ (1.83 m). A very accurate computer model projects a shadow image on the limiting footprint to verify that the robot satisfies the size constraints.

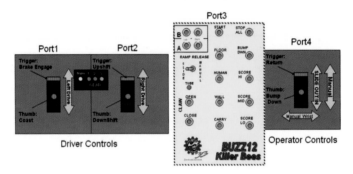

Buzz12 Control Panel

CONSTANTLY UNDER CONTROL

All robot motions were directed with sensor-driven feedback control loops. Operating under continuous feedback, it was determined that it would be difficult to simultaneously control the arm's three degrees of freedom. To automate motion, ten separate closed-loop control algorithms constantly monitored and controlled functions such as the shoulder, elbow, and wrist rotation angle and velocity; the drivetrain speed; and the robot's pitch and yaw rates.

It was intended that the robot continually operate in a form of autonomous control such that the drivers would never have direct manual control of the arm or drivetrain. For example, arm motion was directed as a series of sequential maneuvers based on a prescribed set of conditions for the shoulder and elbow joints. The shoulder was established as the master joint, and its position was used to move the elbow joint relative to the shoulder such that the arm smoothly tracked along a prescribed path. Feedback control produced graceful motion in this dual-joint appendage that would have been nearly impossible to achieve with human directed control.

With the arm functions automated, operator control of the arm was reduced to selecting any of ten push-buttons to place the arm in its starting or stored position, reach tubes along the wall, reach above the player station or on the floor, and score at any level. Once an action was commanded, the robot controller decided how, when, and if it obeyed the requested command based on predetermined internal constraints. Activating each button on the controller directed the arm through a series of positions on its way to the desired destination. The orientation of the tube was maintained with feedback control by coordinating the angular position of the shoulder and arm joints with the wrist position. Feedback control adjusted the wrist angle as the elbow and shoulder extended to keep the tube's orientation fixed at an ideal angle for scoring.

Versatility was provided with the inclusion of a pseudomanual override of the prescribed control functions. When the override was activated, the human operators directed robot functions independent of the preprogrammed, push-buttoned functions. Even these functions were monitored to simultaneously adjust the shoulder and elbow positions and ensure that the commanded actions did not jeopardize the robot's stability or violate any of the system's mechanical limits.

▲ The Killer Bees control panel features ten push-button controls to position the arm and other buttons to manipulate the claw and activate the ramps. Joysticks for driving include buttons to select one of three transmission speeds and a separate joystick offered monitored control of the arm and wrist.

▶ The robot controller simultaneously adjusts the shoulder and elbow positions to keep the tube at a constant height and fixed orientation as it approached the rack. The height is determined to allow the tube to be placed at any point along this path.

Buzz12 Tube Slide Feature

Sophisticated Arm Controls allow Tube to be held at a constant height while being slid in and out.

A single input moves both Elbow and Shoulder Joints in concert to make this possible.

Works on all three scoring levels.

Elbow Motion

Shoulder Motion

Tube Motion

Rack

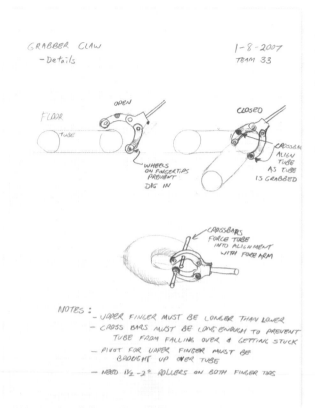

GRABBER CLAW
— Details

1-8-2007
TEAM 33

FLOOR

OPEN

CLOSED

TUBE

WHEELS
ON FINGERTIPS
PREVENT
DIG IN

CROSSBAR
ALIGN
TUBE
AS TUBE
IS GRABBED

CROSSBARS
FORCE TUBE
INTO ALIGNMENT
WITH FORE ARM

NOTES : — UPPER FINGER MUST BE LONGER THAN LOWER
— CROSS BARS MUST BE LONG ENOUGH TO PREVENT
TUBE FROM FALLING OVER & GETTING STUCK
— PIVOT FOR UPPER FINGER MUST BE
BROUGHT UP OVER TUBE
— NEED 1½-2" ROLLERS ON BOTH FINGER TIPS

◀ The first design details for the claw specify parameters for positive control, including a requirement that the upper finger be longer than the lower one, and the addition of crossbars (for stability) and rollers (for picking up tubes) on the fingertips.

Team #33 - Compound Action Claw

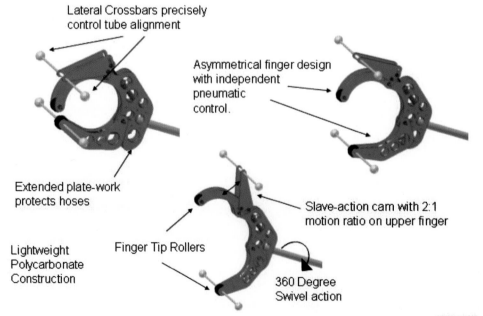

Lateral Crossbars precisely control tube alignment

Asymmetrical finger design with independent pneumatic control.

Extended plate-work protects hoses

Lightweight Polycarbonate Construction

Finger Tip Rollers

Slave-action cam with 2:1 motion ratio on upper finger

360 Degree Swivel action

KillerBees 2007

◀ Tubes are tightly held when the claw is closed. The crossbars provide additional grip on the tube and prevent the tube from twisting.

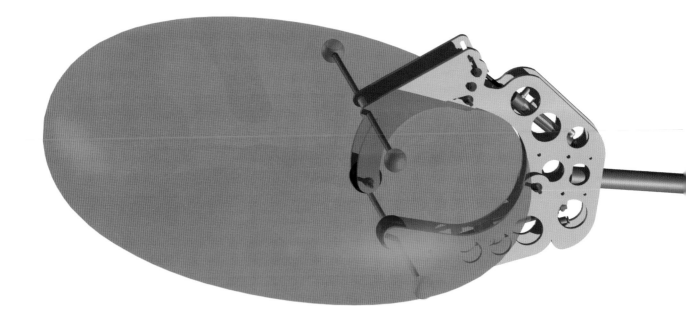

POSITIVE CONTROL FOR POSITIVE RESULTS

Team 33's ability to automatically control every component on the robot greatly enhanced the team's competitive advantage. The robot's claw was one example of the high level of performance made possible by coordinating design and control activities.

Early sketches, completed two days after the game was announced, suggested a claw with two independently operated joints. Crossbars were included on each of the claw's two fingers to precisely control the tube's alignment in the gripping mechanism. The design progressed with an Autodesk Inventor plan for the compound-action claw that included fingertip rollers to ease the claw's movement along the playing field surface. A cam-driven segment of the upper finger provided a range of motion that reached over the tube before clamping down on the tube in the claw's lower thumb for a secure grasp.

For retrieval of a tube from the floor, a control algorithm dictated the sequential operation of each of the two pneumatic pistons. One-touch control automatically oriented the claw at a precise angle relative to the floor. When a tube was contacted, pressing the close button on the driver's station caused the upper finger to close. This action trapped the tube and the claw's crossbar aligned

the tube with the claw. A tenth of a second later, the lower thumb closed to grab the tube while lifting it off the floor. A similar set of prescribed movements directed motion to grab a tube leaning on the playing field wall.

Expert control was also practiced during the autonomous period. At the start of each match, the three joints in the arm and wrist were activated to precisely position the tube for scoring. The robot moved forward along a straight path based on feedback from wheel encoders that measured speed and a gyroscope that measured the robot's direction. Once the camera acquired the target light, the robot moved toward the goal using the camera's tilt angle as a measure of the distance to the target.

The robot precisely measured its speed, position, and heading angle as it advanced, and it decelerated to a stop as it neared the goal. The arm and claw were then automatically operated to score the tube. Upon scoring, the robot controller applied the recorded path of motion to automatically navigate back to the player station to position the arm to receive a tube at the beginning of the teleoperated period of play. This entire sequence was accomplished in as little as four seconds—a speed that would be hard to match with human control.

THE VALUE IN PRIORITIZING CONTROL

The most amazing accomplishment of Team 33 was their ability to develop their control system before the robot was constructed. By using a wide variety of computer programs, the team analyzed the robot's motions and developed paths of motion that the components needed to follow. The computer analysis included determining the output from the robot's sensors as the components moved in the prescribed sequences. This information was tabulated as a series of required positions for each robot motion with the tables serving as the input commands for the physical robot. When the actual robot was constructed, the tabulated values were compared to the actual sensor measurements and feedback control was used to bring the two values in agreement.

The Killer Bees focused on effective control from the very start of the design process. Components that favored positive control were designed and fabricated, with sensors embedded in the design to constantly monitor robot functions. Continually operating under automatic control resulted in accurate motion and allowed the drivers to concentrate on playing the game rather than manipulating the robot.

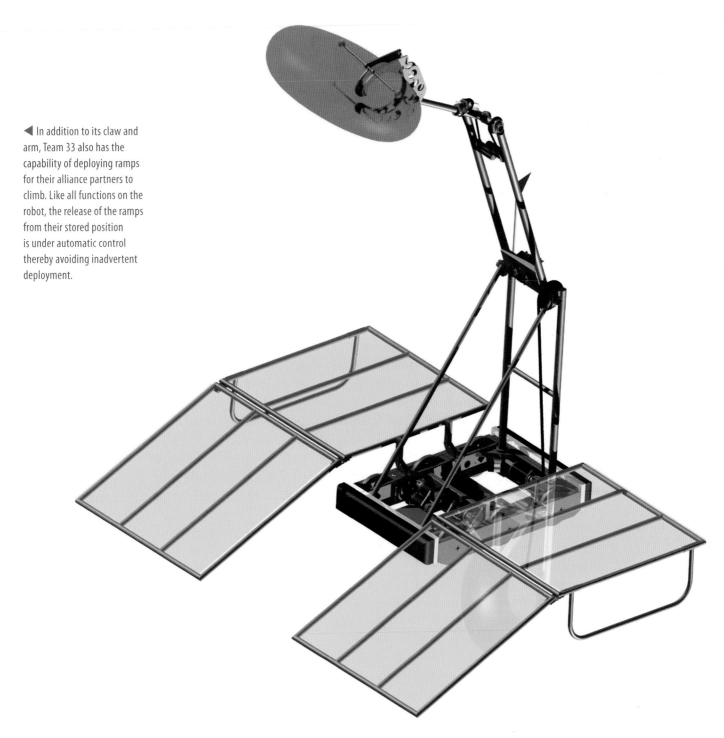

◀ In addition to its claw and arm, Team 33 also has the capability of deploying ramps for their alliance partners to climb. Like all functions on the robot, the release of the ramps from their stored position is under automatic control thereby avoiding inadvertent deployment.

Ease of Control with easyC Programming

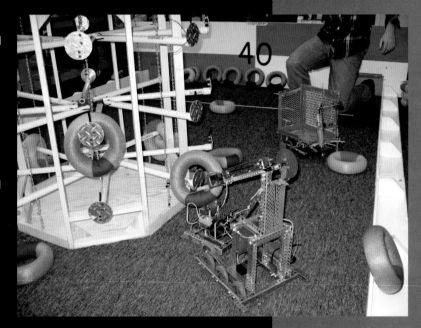

▼ A scale model version of Rack 'n' Roll was created to test game strategies and robot designs. All elements of the field, rack, and game pieces were appropriately scaled to closely mimic actual play.

Team 40, based in Manchester, New Hampshire, begins the robot design process by thinking small before going big. While many robot designers prototype mechanisms prior to building the final devices, Team 40 prototypes strategy.

To investigate potential game strategies, a ⅓ -scale model of the game elements was first constructed. Similarly scaled robots were created as physical realizations of the team's brainstorming ideas. By playing the game with these miniature prototypes, the nuances of the game were revealed and a winning strategy was discovered.

To test ideas, minibots were constructed using the Vex Robotics Design System and programmed using easyC PRO programming software, a graphics-based programming language for robot controllers. Programming expertise is guaranteed on Team 40 since the team's sponsor, intelitek, produces the easyC PRO programming software.

The miniaturized game convinced Team 40 to build a tube-scoring robot. They decided this design would control its own destiny and not have to depend on the capabilities of its alliance partners. To counteract alliances that could score sixty points with two robots on ramps, Team 40 reasoned that a tube-bot would have to be fast and efficient. To score long rows of tubes, an integrated system of mechanisms, sensors, control, and skill was needed. It was predicted that a winning tube-scoring robot would have many of its functions automated to minimize tube manipulation time and maximize scoring.

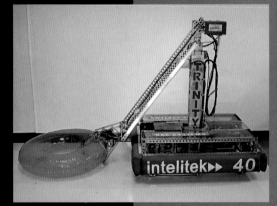

▶ The claw snaps into position at the end of the arm and closes around a tube. The tube's horizontal orientation is maintained as the arm rotates by the four-bar linkage that connects the claw and the elevating mast.

Small-scale testing indicated that ringers were more easily scored when they were kept parallel to the ground and pulled down over the end of the spider leg. It was also discovered that robots were more apt to become entangled in the rack when they were close to it while scoring, so it was critical that they maintain a safe distance from the rack. A third key observation from small-scale play was the necessity to score on all three levels of the rack.

These findings shaped the design of Team 40's full-size robot. A hinged claw snapped into place to grab tubes from the floor. A four-bar linkage connected the claw to the elevating mast and maintained the tube's horizontal orientation as the arm rotated. The linkage was long enough to keep the robot away from the rack when scoring. The three-stage elevator enabled scoring on all three levels. For maneuvering near the rack, the scoring assembly was mounted on a turret capable of rotating 720 degrees. A system of sensors and controls transformed these parts into the quick scoring tube-bot the team first envisioned during their small-scale prototype tests.

▲ Team 40's robot features a fast-acting claw, a rotating arm that fixes the tube's orientation, and a telescoping elevator mounted on a rotating turret.

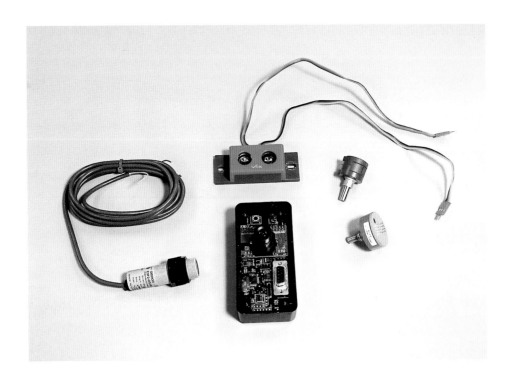

A SUITE OF SENSORS

The need to quickly grab and score tubes necessitated automating many of the robot's functions. As such, the robot's control functions were designed in parallel with the mechanical systems and a series of sensors was incorporated in the design to monitor robot conditions. To ensure compatibility between the sensors and the systems, CAD drawings of the robot components included the sensors. Such advanced planning secured the sensor's placement and detailed the mounting mechanism required for each sensor. Similar to the resemblance between the miniature prototype and the final robot, careful CAD design produced accurate plans to construct the functional components.

A variety of sensors monitored functions and provided feedback to the robot operators. A proximity sensor was mounted behind a viewing portal cut into the palm of the claw. This sensor emitted infrared light and detected when that light was reflected back by a tube. Once a tube was detected to be within reach, LEDs mounted along the length of the arm were automatically energized to alert the drivers. With this system, the drivers' visibility was expanded and they could easily pick up tubes when their direct view of the claw was blocked.

One of the most sophisticated sensors was the CMU camera to track the vision targets mounted on the rack. When the camera saw the green target light, the vertical and horizontal locations of the light within the camera's field of view were relayed to the robot controller. The vertical location correlated with the distance from the rack, and the robot controller used this signal to drive the robot toward the vision target. The horizontal location indicated the rack's orientation, and this signal was used to rotate the turret into scoring position. An ultrasonic sensor, mounted at the front of the robot, served as a back-up sensor to determine the distance to the rack.

Additional sensors monitored other robot functions. Encoders were mounted on each side of the drive system to measure wheel revolutions. These signals were the input to a control algorithm to drive the robot. When the robot was pushed, the drivetrain automatically pushed back to maintain position. The sensors were also used for straight, accurate driving. Potentiometers measured the turret orientation, elevator height (by measuring the rotation of the elevator lift motor), and the arm angle. These sensors measured every powered system on the robot. And because each function was measured, each function was controlled.

▲ Sensors monitor all robot functions. An infrared sensor detects tubes ready to be grasped; an ultrasonic sensor measures the distance to the rack; a camera tracks the target light on the goal; potentiometers record the position of the arm, elevator, and turret; and encoders monitor the rotation of each drive wheel.

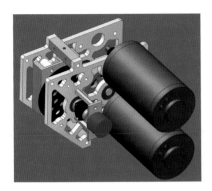

◀ A potentiometer provides feedback on the lift mechanism for the robot's mast. Mounting the potentiometer on the gearbox output shaft provides the highest degree of resolution for the lift.

▼ The infrared sensor emits its signal through an opening cut into the claw's back plate. Rubber material on the bottom of the claw improves the claw's grip.

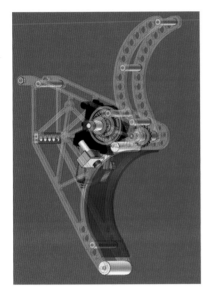

◀ An infrared proximity sensor is integrated in the design of the claw. The sensor indicates when a tube is ready to be pinched by the claw and triggers the LEDs on the arm to illuminate.

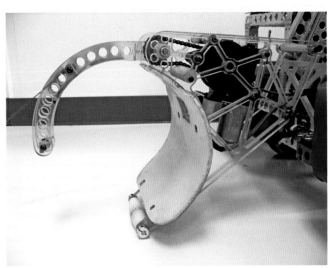

▶ The infrared proximity sensor illuminates LEDs on the arm to signal the drivers that a tube is within reach. The lights are bright enough to shine through the tubes and greatly aid the operators' ability to retrieve ringers on the far side of the field.

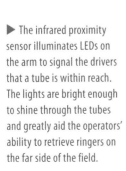

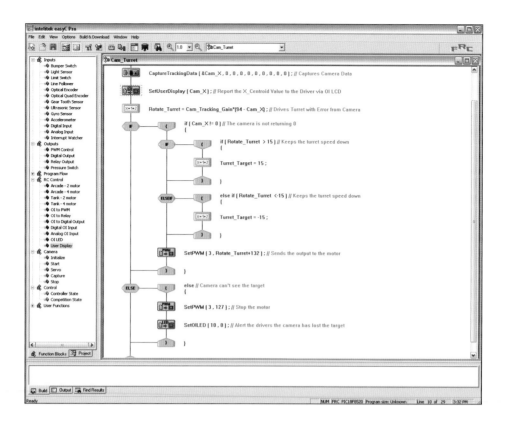

AN INTEGRATED SYSTEM OF SENSORS AND CONTROL

Computer programs were written to transform the sensor outputs into robot functions. Using a computer program to write code for the robot control is a natural step for Team 40, given their sponsor is the author of easyC PRO, a graphical programming language that uses icons to represent measured inputs, controlled outputs, and computer subroutines.

Graphical programming is an intuitive programming method. Creating computer code in this environment consists of dragging an icon into a window, linking that icon with other icons, and downloading the resulting program to the robot controller. To help the user understand the computer code that is generated when icons are linked together, easyC PRO lists the actual C-programming code next to the icon. This feature provides programmers with insight into the generated code, which can be a tool to learn text-based C-programming.

Team 40 programmed their robot with easyC PRO to automatically perform robot functions, including autonomous control. A collection of autonomous programs was written to score from any starting position on either of the lower two levels of the rack. Switches on the robot were set to select the specific autonomous program for each match.

A diagnostic program was written to view the feedback from each sensor, with the resulting information displayed on a computer terminal window using an easyC PRO subroutine. By including this diagnostic program as one of the selectable autonomous programs, Team 40 could quickly evaluate each sensor without having to download a separate testing program. The program converted the measured voltage signals to the appropriate units for each monitored system. For example, the turret and arm positions were displayed as angles, and the elevator mast height was displayed in inches. The terminal window diagnostic program was a useful troubleshooting tool to monitor the sensors.

▲ The camera-tracking code is developed by linking a series of icons in intelitek's easyC PRO programming software. This graphical programming environment includes subroutines that are combined with user-written code.

▲ Autonomously scoring a tube is a challenging task. Team 40 successfully applies data from an onboard camera to locate the goal, drive the robot to scoring position, and rotate the turret to place the tube over the end of the spider leg to score.

▶ A diagnostic program to verify feedback from every sensor runs in the background and uses a graphical display feature in easyC-PRO to display the results.

VIDEO GAME CONTROL OF THE ROBOT

Accurate driving was important, and Team 40 reasoned that drivers who are comfortable with maneuvering their robot are more apt to use the robot to its highest potential. Given their students' familiarity with video game controllers, Team 40 used the popular Xbox control interface to drive and operate the robot. One controller commanded actions of the turret, claw, and elevator. The second controller operated the drive system. Additional control functions were mapped to switches on each handheld controller. A commercial interface, the USB-Chicklet, converted the USB Xbox signal into a form that could be processed by the operator interface.

To improve the efficiency for retrieving tubes and scoring, computer algorithms were written to maneuver the arm to preset conditions and execute a series of commands to place tubes on the rack. Four maneuvers were programmed and mapped to buttons on the Xbox controller to position the claw in the pickup position or at one of three heights for scoring. With a tube in position above a spider leg, the operator resumed manual control to drop the elevator and score the ringer on the rack. The preprogrammed maneuvers simplified the robot's operation and greatly improved scoring ability.

Because the turret rotated 720 degrees, tubes could be picked up from either the front or rear of the robot. With a tube in hand, the operator drove the robot straight to the goal and did not have to spin the robot around to correctly orient the robot with the goal. To ease the transition associated with driving a robot in reverse, a special feature was programmed and activated using a button on the Xbox controller. Termed bidirectional drive control, the driver could push one button on the Xbox controller to automatically rotate the turret 180 degrees and invert the driving direction. This ability to retrieve and place tubes without having to spin the robot greatly increased the team's scoring efficiency.

◀ The USB-Chicklet is an interface between the Xbox 360 controller and the robot controller. The resulting hand-held control is a form very familiar to robot drivers.

▼ All functions on the robot are easily controlled using two Xbox 360 controllers: one for driving and the other for manipulating tubes.

Arm Operator

Driver Control

DESIGNING WITH CONTROL IN MIND

Team 40's initial work with the miniature-scale version of the competition was an accurate forecast of the full-sized game. The scale-model provided the team with a forum to rapidly evaluate strategy and robot designs. This activity accurately predicted important aspects of the game and inspired essential features for the full-sized robot.

The actual game was indeed very fast paced and the most successful robots were those that could score quickly. From design to competition, Team 40's primary goal was efficient robot control. In the end, an integrated control system of sensors, handheld controllers, and clever programming proved to be a winning combination for Team 40.

▼ An integrated system of mechanisms, sensors, programs, and driver control provides the capability to quickly pick up tubes and score them on any level.

Making Control Look Easy

Receiving the Rockwell Innovation in Control Award was no shock to Team 386, which claimed that winning was easy. This perspective stemmed from the fact that the most important sensor on the robot was the Staples Easy button, often featured in advertising as a single magical button to solve all kinds of problems.

The Easy button was a critical sensor on the team's robot. When it was activated, it signaled that a tube was scored and all was indeed well. Upon scoring, a team member's cheer of "That was" would be answered immediately by the rest of the team saying, "easy."

Early analysis of the game forecasted that the most problematic element would be placing a ringer on the scoring rack. Given this challenge, the team set out to make this task as easy as possible for the drivers. Team 386's discovery of the easy method for control resulted from the team's insistence that feedback should be an important design feature. The team's control elegance involved an innovative method of creating a vacuum, an integrated collection of sensors, and the Staples Easy button.

▲ The Easy button on the Team 386 control board advertises the team's claim that control is easy. The control station also features indicator lights to signal vacuum status, selectable positions for the robot arm, and triggers to align and rotate tubes.

▶ Team 386 cleverly positions a switch that is activated by the rack when a captured tube is in scoring position. The red suction cups mounted at the end of a Y-shaped yoke fall on a tube, with a vacuum drawn to gain control of the captured tube.

DISCOVERING HOW TO PULL A VACUUM

An analysis of scoring methods determined that a vertical tube orientation was desirable. Since the spider legs moved, the round edges of the tube naturally centered the tube on the leg. If the tube was held at the correct height, an operator only needed to drive the robot straight into the spider leg to score. Critical concerns in this scenario were securely holding tubes and avoiding getting tangled in the rack when placing tubes on the goal. These concerns prompted investigating a vacuum system to hold the tubes.

Design goals were established to evaluate vacuum-generating systems. The system had to create sufficient vacuum to lift a tube off the ground and hold it until it was scored. The system also had to alert the operators if a proper vacuum was not achieved. Finally, the robot had to be able to start a match holding a tube. These goals provided a framework for examining the potential to use the venturi vacuum provided in the Kit of Parts or construct a new vacuum system.

Testing indicated that the venturi vacuum needed a constant supply of air to operate, thereby forcing the robot's compressor to continually run and draw from the battery. Also, this system could not hold a tube before a match. The team investigated a motor-driven vacuum pump, but the requirement to hold a tube at the start of the match negated the value of this solution.

A workable vacuum method was designed with two pistons in series, connected by their output shafts and mounted opposing each other. When compressed air expanded the master piston, the other piston retracted. As the second piston, called the slave piston, retracted, air was drawn into its output port. When the output port of the slave piston was closed, the retraction created a vacuum.

The slave piston's output port was connected to a suction cup to grip tubes. Two identical vacuum systems were designed, each connected to a separate suction cup to create redundancy in the system. The suction cups were mounted at the end of a Y-shaped yoke that was lowered on top of a tube.

The travel distance of the master cylinder shaft was monitored to provide feedback. If the slave cylinder did not resist being expanded, for example when the output port was open, the master cylinder could retract its full length. When the master cylinder did not fully retract, the slave cylinder was pulling a vacuum.

Cylinders with a magnetic bore plate were used as the master pistons. Reed switches—small switches that close in the presence of a magnetic field—were mounted on the outside of the slave pistons at the end of the stroke. If vacuum was not achieved, the master cylinders traveled their entire stroke length and activated the reed switches.

If the suction cup was not in strong contact with a tube, a vacuum was not achieved. Lights on the operator panel relayed the status of the vacuum, with green LEDs signaling when vacuum was achieved and red LEDs indicating there was no vacuum. This feedback signal also initiated a programming function to automatically reactivate the vacuum system. The master cylinder was quickly cycled three times if vacuum was attempted but not achieved. A failure to obtain vacuum after this automatic cycling required the suction cups be repositioned.

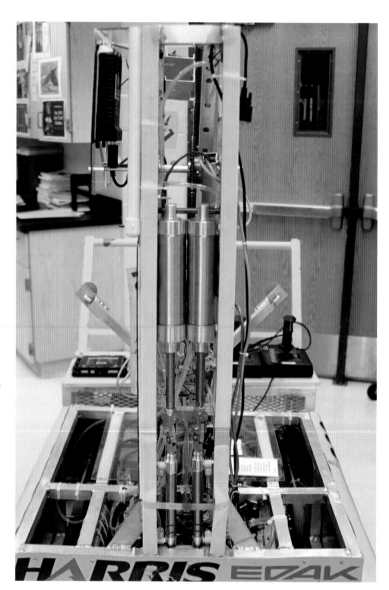

IMPROVING THE VACUUM SYSTEM

Testing revealed two problems with the original design: inadequate suction cups and the need to align the tube with the yoke. Team 386 began the competition season using the suction cups provided in the Kit of Parts. The rigidity of these devices required that they be placed along the very top of the tube to create a seal. These cups could only obtain suction, on average, about once every three times they were placed on a tube.

▲ Smaller pistons act as the master cylinders and retract to draw a vacuum on the larger slave cylinders. The robot is equipped with separate vacuum systems for each suction cup.

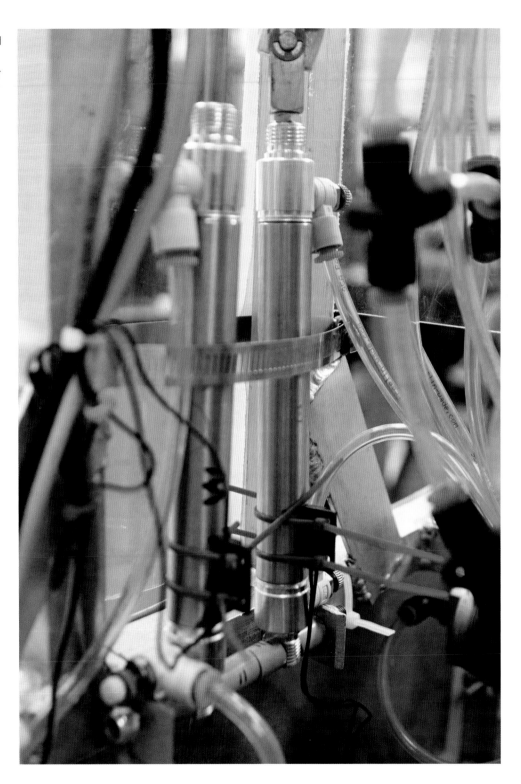

▶ Reed switches are attached to the lower ends of the master cylinders. If the master cylinder is retracted its full length, the reed switch closes and indicates that a vacuum was not obtained.

At the initial competition, Team Voltage noticed that Team 230 had a much higher success rate retrieving tubes. Discussions with Team 230 prompted Team 386 to explore other suction cup options. A review of other possible suction cups followed, with a bellows-based design proving the most effective. With the assistance of another team and by changing two parts of the robot, Team Voltage's ability to pick up tubes shot up to 80 percent at subsequent competitions.

Testing revealed that if a tube was not centered in front of the robot, it was difficult to pick up. To correct this, a tube-alignment device was added to the robot—a pliable Lexan sheet stretched across the front of the robot, with its ends connected to pistons. To center the tube between the suction cups, the operators extended the alignment device when in proximity of a tube. Once centered, the yoke was lowered and the suctions cups were ideally located along the tube's upper edge.

TUBE POSITION CONTROL AND THE EUREKA MOMENT

Autodesk Inventor software was used to evaluate the robot geometry required to place tubes on the rack. Once suction was obtained, the yoke was rotated 90 degrees to hold the tube vertically. The CAD drawings were instrumental to design an arm that retained this orientation as the arm rotated to reach all levels of the rack.

▲ A variety of suction cup alternatives are tested before determining that a bellows design has the best ability to maintain suction over the largest surface area on the tubes.

◀ A retractable tube-aligning device centers the tube such that the suction cups are properly placed on the centerline of the tube's upper surface.

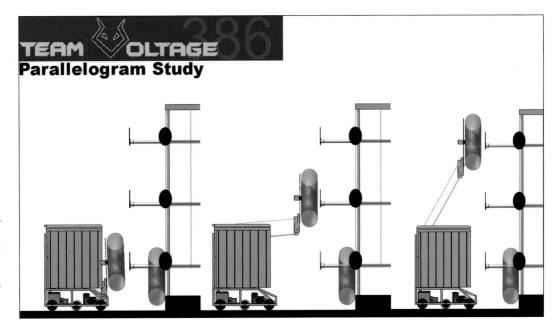

TEAM VOLTAGE 386
Parallelogram Study

▶ An Autodesk Inventor model examines orientations to place a tube on each level of the rack. The alignment between the center of the tube-holding device and the end of the spider leg prompted the idea to mount a button in this location to indicate when the tube is ready to be released.

A review of these drawings revealed that once a tube was in scoring position the end plate of the tube aligned with the center of the yoke. The original plans called for the operator to drive forward a sufficient distance such that the second operator could release the tube. This action required close communication between the operators, often when their view of the robot was blocked.

Since the yoke was centered on the spider leg, a switch on the yoke could indicate when contact with the end plate was made and it was safe to release the tube. After this switch was added to the CAD drawing, a team member casually mentioned that the switch looked like the famous Staples Easy button. That comment led to using the Easy button on the yoke.

Further testing verified the correct placement of the Easy button on the yoke so that it was activated when a tube was in scoring position. When activated, the tube was automatically released and the robot backed away from the rack. Because the button was activated by the rack, it was only pushed when the tube was on the rack and could be safely released. The placement of this important sensor made scoring on any level easy.

MORE SENSORS LEAD TO BETTER CONTROL

The angle of the robot arm was measured with an optical encoder. Preset arm conditions were determined and correlated with encoder measurements at each of the three desired locations. Control algorithms positioned the arm at the correct angle. To reset the encoder, a limit switch was mounted on the arm and activated when the arm was in its lowest position.

▶ The Easy button is mounted in the center of the tube to contact the end-plate on the spider leg. Its activation automatically releases the tube on the rack.

▼ Linkages on the arm assure that the Easy button maintains its vertical orientation as the arm rotates. The offset distance from the front of each tube is an important dimension to guarantee the button activates when the tube was well beyond the end of the spider leg and in position to be scored.

◀ An optical encoder is connected to the shaft that rotates the arm. The encoder measures the arm's angle of rotation, and control algorithms accurately position the arm at each of the rack's three scoring levels.

▶ When the arm is in its lowest position, a limit switch mounted on the arm comes in contact with the robot chassis. This signal establishes the

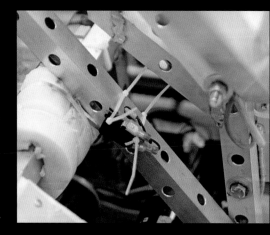

The CMU Cam2 camera was mounted at the top of the robot and protected with a transparent shield. In autonomous mode, the camera's signal drove the robot to the nearest signal light on the goal. The camera also controlled a servo motor to tilt the camera and track the light. This tracking signal was used to determine the distance to the goal. In autonomous mode, the camera drove the robot forward until the Easy button was activated, causing the release of the tube on the goal.

Team 386 created a machine with a smart vision system capable of autonomous scoring, an innovative method to create a vacuum, and a strategically placed button to enable automatic scoring. Team Voltage's consideration of controls from the very beginning of the design process led to a highly effective machine that ultimately made scoring look easy.

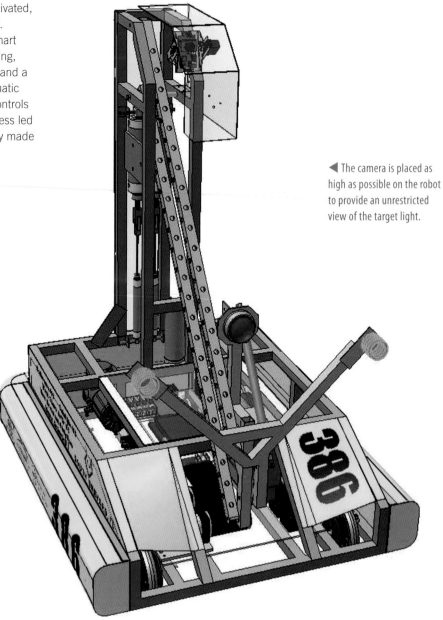

◀ The camera is placed as high as possible on the robot to provide an unrestricted view of the target light.

▲ For Team 386, easy control resulted in a high performing robot that earned the team the Rockwell Innovation in Controls Award at the 2007 FIRST Championship.

Sophisticated Systems for Successful Control

The double-articulated arm on Team 418's robot, sponsored by National Instruments and BAE Systems, significantly influenced the control system design. Two means of control were initially proposed: independent direct control of each joint and relational control to manipulate the two joints in tandem. A functional test determined the best control system to expertly position the arm.

Two motors powered the arm, with a larger gear ratio used at the lower joint to handle the higher loads. The tube manipulator, fixed in position at the end of the arm, used two additional motors attached to rollers capable of drawing in and pushing out tubes. Since scoring required mastering these four motors, an efficient control system was required.

Direct control was first evaluated using the completed robot. Within minutes of the test, the operator drove the arm into the floor and damaged the manipulator. The operator was easily confused trying to determine which joystick controlled which motor, and moving each joystick in the proper direction. The task was especially troubling because the robot was capable of grabbing tubes from both the front and back of the robot, thereby negating the use of any operator memory tricks based on moving the arm up or down. Mastering direct control would require weeks of operator training.

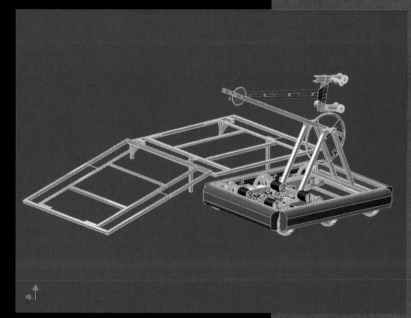

▲ Team 418's robot includes a ramp and a double-articulated arm. Using the same transmissions for the six motors that powered the arm and drive systems increased reliability and eased maintenance.

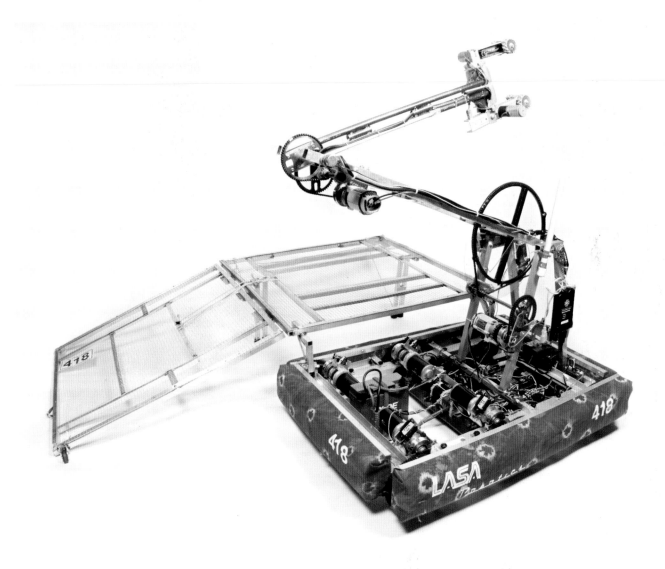

DAMAGE CONTROL

While the robot was being built, the control team developed and programmed a relational control system for the arm and manipulator. This integrated system of hardware and software resulted in an intuitive system for arm control. Using this control, the operator was able to pick and place tubes on the rack within minutes of its use, and physical stops on the controller prevented the robot arm from impacting the floor and damaging the manipulator. This sophisticated system for successful control averted ungainly movement and provided inherent protection to the mechanism.

▲ The folding ramp provides a final parking spot for an alliance partner at the end of the match.

COMMANDING RELATIONAL CONTROL

Named after inventor Lewis Henrichs, a high school sophomore on the control team, the Lewis-Arm was a double-articulated ⅛-scale model of the robot arm. Rotary potentiometers at each joint on the miniature arm were calibrated to perform like the potentiometers measuring the robot arm's position. The Lewis-Arm was mounted on the driver control station such that the physical stops kept the robot arm within a safe operating region.

Signals from the controller's potentiometers were compared to those produced by the robot's potentiometers to determine if the robot's arm orientation matched that described by the controlling arm. If the potentiometer readings did not match, the motor powering the unaligned joint was signaled to align the joints with the commanded positions. Because each linkage moved in relation to the other, the resulting control signals were similarly related.

Because of the arm's mass and speed of motion, the arm had considerable inertia. To account for the arm's desire to continue to move after motor power stopped, the control algorithm applied a significant dead band near the targeted position. Rather than waiting for the arm to reach the exact targeted location to stop the motor, power was cut off once the arm entered the prescribed dead band. Using this control strategy, the arm would then coast into its final targeted position. This algorithm moved the arm quickly but was not precise. Precise arm control, if needed, was achieved by activating a button that switched the control algorithm to direct control.

Two small wheels, each independently powered, were mounted on a U-shaped bracket at the end of the robot arm to manipulate tubes. When the wheels rotated in opposite directions, tubes were picked up or released, depending on the rotational spin of the wheels. When the wheels were spun in opposite directions, the tube was rotated while held by the manipulator—a feature that allowed for optimal scoring orientation. A joystick handle served as the outer linkage on the miniature arm. The joystick's finger and thumb buttons controlled the tube manipulator motors to fully control all aspects of the robot's arm and manipulator from the Lewis-Arm.

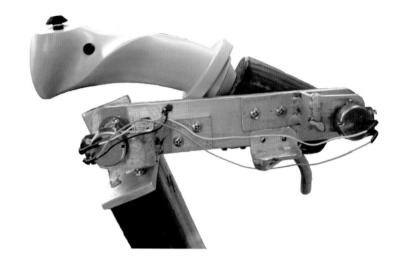

▲ The miniature control arm includes rotary potentiometers to record the arm's location and physical stops to prevent the arm from extending beyond a safe operating range. Joystick switches command the tube manipulator.

▶ Rotary potentiometers at the robot's joints measure the position of each linkage. The orientation of the tube is determined by two sets of wheels at the end of the manipulator that are controlled to hold and position the tube.

INTEGRATING HARDWARE AND SOFTWARE FOR EXPERT CONTROL

Team 418's operator interface was an integrated system of student-written software and sophisticated data processing hardware. Robot sensor data collected by the IFI control system was repackaged in the IFI operator interface before being exported. The exported data was processed on a laptop computer using National Instruments LabVIEW software. The signals were analyzed using LabVIEW—a graphical programming language for acquiring, processing, and displaying measured data—to create motor-control commands that were exported from the laptop and transmitted to the robot.

A dashboard display on the laptop monitor relayed the condition of each robot system. The robot's arm position was indicated as a graphical image placed in front of an image of the miniature control arm. This real-time representation of the arm's position tracked the arm's ability to reach its commanded position. The display also indicated when the arm reached each of the three scoring levels to increase the driver's ability to place tubes on the rack. This same information was also displayed using three LED lights mounted on a tower next to the miniature arm. To signal when the arm was dangerously close to the ground or being operated at a condition that could cause the arm motors to overheat, all three LEDs flashed to alert the driver.

The laptop dashboard also displayed the status of the primary and back-up batteries, the drive motors' speed, and output signals from gear tooth sensors that monitored the drive system's transmissions. The status of the motors powering the arm was measured with current sensors that signal if the motors were stalled and overheating. The graphical display provided a quick and accurate method for the operators to monitor the robot's performance.

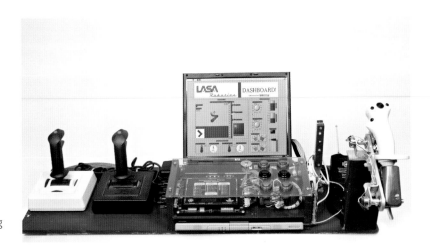

▲ The control station consists of joysticks for driving the robot and operating the miniature arm. Control signals are processed using the IFI control system and a team-written LabVIEW program that runs off a laptop computer.

▼ A dashboard presents information on all of the robot's systems. This display is created with LabVIEW to provide a visual means to monitor the robot's performance.

TESTING THE CONTROL SYSTEM

A National Instruments data-acquisition device was used with LabVIEW to understand how each sensor worked prior to being installed on the robot. A LabVIEW program was run to display an oscilloscope monitor to indicate the signals produced from each sensor. Once understood, the sensors were connected to the IFI robot controller, and a custom LabVIEW dashboard program was written to indicate how the controller interpreted data from the sensors.

Team 418's control algorithms were also evaluated using a LabVIEW graphical program. For example, potentiometers were directly connected to the robot controller and operator interface to simulate commands from the miniature arm and the position of the robot joints. The graphical output validated that the correct signals were generated to drive the motors. This type of bench-top testing was a valuable troubleshooting technique to evaluate the code before the robot was constructed. Control algorithms were also tested on a simple drive platform that the team had previously constructed. By developing and testing the code as the robot was constructed, the control software was ready when the robot was completed.

▶ The IFI robot controller and operator interface are used as a test set-up to investigate possible control sequences. Sensors are used to input data to the system and the created output signals are evaluated to test the control system's accuracy.

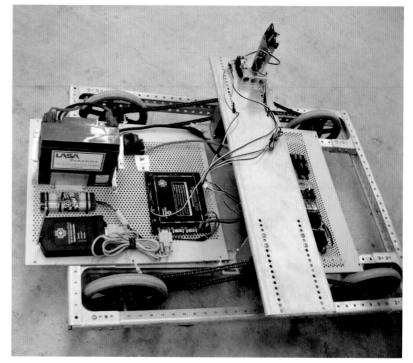

▶ A simple drive platform models the competition robot's drive system and serves as a tool for troubleshooting control programs.

◀ Modular construction allows components and systems to be constructed at the same time and then integrated together. The electronics module is developed on a platform and is fully tested before being mounted to the robot base.

▶ Structural components of the arm are added to the base even before the drive motors are installed. The availability of the separate drive platform eases the need to have the competition robot available as a control system test platform.

▲ ▶ The final robot is a tightly packed, highly integrated system of components. The large sprockets on each arm joint reduce the joint's speed while increasing the available torque.

MONITORING ALL ROBOT FUNCTIONS

Additional sensors and control function were used to monitor and affect other systems. A three-speed drive system afforded a range of driving options as commanded by the driver using the top buttons on the drive joysticks. To enhance maneuvering at high speed, a smart-shifting algorithm was created to automatically downshift the drive motors while turning. Once the turn was completed, the controller would reset the drive motors in the higher gear to resume high-speed motion. The gear setting was displayed on the dashboard to inform the drivers.

As another example of coordinated control, a two-step ramp release mechanism was used to avoid release before the end of the match. To drop the ramp, the robot driver armed the ramp by holding down both buttons on the drive joysticks, a command that would only be initiated when the robot was not moving. Once the system was armed, the robot operator pushed a ramp release button that caused a motor to pull a retaining pin that released the ramp.

Expert control of Team 418's robot was achieved by carefully developing the control programs while the robot was being constructed. Using LabVIEW, the control program was tested and debugged independent of the robot mechanisms. Coordination between the build and control teams ensured that the integrated system of hardware and software expertly executed the operators' commands.

▲ Team 418's success resulted from a coordinated team effort between the design, manufacturing, and control subgroups.

System Simplicity Succeeds

Team 1629's 2007 FIRST Robotics Competition robot can be summarized as two major systems for two methods to score. The primary scoring system was a single-joint arm that was lifted on a belt-drive elevator to reach all scoring levels. The secondary scoring system was a ramp and platform capable of providing an end-of-match parking place for two alliance robots. Integrating these two systems required creative design and effective control, aided by the simplicity of each system.

Simplicity was a guiding principle for Team 1629. It was reasoned that a fewer number of moving parts required fewer control functions, thereby simplifying the tasks of the team's mechanical and software developers. A simple system would also be easier to fabricate, leaving more time for testing and driver training before the competition. The resulting robot consisted of a tower, arm, claw, and ramp. In addition to the drive system, only three motors powered these systems, using a belt drive for the elevator, a rotary lead-screw for the arm, and a small motor for the claw. With only three motors to monitor all these functions, the control system, by design, was simple.

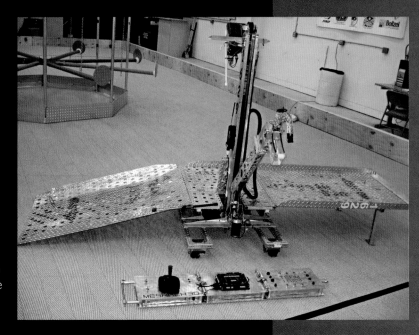

▲ Team 1629's robot includes two highly reliable scoring systems. A single-joint arm and claw ride a central elevator to place tubes, and a ramp platform provides a means for alliance partners to earn bonus points.

OPEN AND CLOSED LOOP CONTROL

Open and closed loop control was applied to manipulate the robot. When operating with open loop control, the robot drivers manually positioned the robot and its systems. In closed loop control, the robot controller used feedback from a camera, gyroscope, potentiometers, and limit switches to direct robot functions according to the team's programmed instructions. The dual systems of control provided for fail-safe operation and allowed the robot operators to choose the control method they were most comfortable with to enhance the quality of the human-robot interaction.

Closed loop control routines simultaneously maneuvered the elevator to a predefined height while the arm was adjusted to an optimal angle at each scoring level. An additional control placed the arm in the required location to retrieve tubes from the floor. To initiate closed loop control, the robot operator selected any of four conditions using push buttons on the operator control board. Additional automated functions included a sequenced process to carefully open the claw when scoring so that its opening speed did not dislodge the tube from the spider leg. Open loop control was provided with toggle switches for control options beyond those available with closed loop control.

The robot drive system also operated under closed loop control by following a programmed path plan during autonomous mode. A yaw rate gyro monitored the robot's heading and its input guided movement along a straight path. A camera tracked the target light and once the target was acquired, this signal directed the robot to the goal. In open loop control, the driver directly commanded the robot's drive system using a single joystick. To drive the robot perfectly straight, the gyro could be used to augment human-directed drive commands.

The simplicity of the robot design required simple control functions. These simplified controls allowed the operator control station to be relatively uncomplicated and highly reliable. The control station was constructed with a minimal number of switches and indicators, thereby easing the board's design and construction.

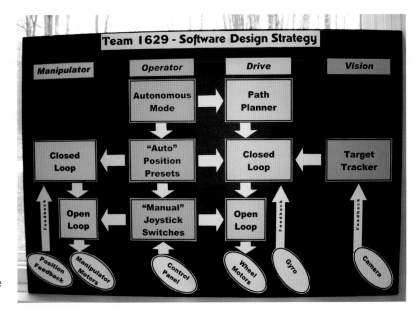

▲ Sensor data and operator commands are used separately and in parallel to create open and closed loop control strategies.

▶ A combination of preset switches, manual control switches, and feedback indicators afford the arm driver a range of options for controlling the arm and claw.

▶ The base driver's controls include a joystick to maneuver the base, a switch to engage the gyro, and indicator lights to signal the status of the gyro and camera.

ARM AND CLAW CONTROL

A belt-driven elevator housed a carriage that supported the robot's arm. The elevator was powered from a motor located in the robot's base, with the drive belt running from this motor to the top of the elevator's mast. Each end of the belt was connected to the arm carriage to pull the arm to the commanded position. A ten-turn potentiometer was attached to the belt sprocket at the top of the mast to measure the position of the carriage as it moved along the elevator's mast.

The arm was positioned with a motor driven lead-screw. With the screw's threads attached to the arm, the arm pivoted when the lead screw rotated. A rotary potentiometer was mounted at the arm's pivot point to measure the system's rotation angle.

The elevator and arm potentiometers provided an accurate means to measure the arm's height and angle. To zero each potentiometer and establish a reference position, two push buttons were located on the driver control panel. When these buttons were pressed at the same time, the current position of the sensors was established as the home position. This feature allowed the home positions to be updated quickly and easily, while the double-button protection prevented the positions from being set accidentally.

Closed loop control applied data from these sensors to automatically position the arm at predetermined locations for retrieving and scoring tubes. A generous dead band was specified for the elevator height control system to prevent this system's motor from overuse, as would result if the height was narrowly controlled. Limit switches were also installed near the ends of the elevator path as a secondary means to monitor the carriage position. Sensor data helped reduce the elevator's speed as it neared its endpoints to prevent the system from overshooting its commanded position and colliding with mechanical limits.

The present conditions allowed the operators to accurately position the arm by merely pushing

a button. This automatic control feature positioned the arm faster and more accurately than if controlled by a human operator. Placing the arm in an exact location was especially important when retrieving tubes, since an incorrect arm angle resulted in an undesirable orientation of the captured tube and prevented easy scoring.

Rather than establishing the preset arm conditions as specific lines of instruction in the control software, the commanded height and angle could be specified in the software by physically positioning the arm in a desired orientation. A software routine recorded these values and the control program referenced these saved positions when the preset commands were initiated. If the drivers preferred a different tube-placement orientation, the robot was placed in that position and the robot was commanded to learn the new position. This ability increased the system's reliability as the actual code was not changed by a programmer each time a new position was desired, thereby minimizing the time and process to adjust the control system.

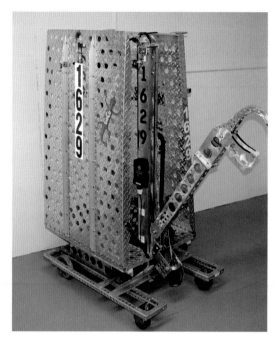

◀ A timing belt runs from the base of the robot to the top of the center mast. The belt is attached to the base of the arm, with a carriage riding up and down the mast to position the arm at specified heights to retrieve tubes and score them on the rack.

▶ The drive motor for the elevator is located at the base of the mast and directly connected to the belt sprocket. The arm angle is positioned using a motor-driven lead screw mounted at the base of the arm.

▶ A potentiometer is mounted to the upper sprocket on the elevator belt drive to determine the arm's height. The camera, housed in a protective enclosure, is also mounted at the top of the mast to provide a direct line of sight to the target light.

POSITIVE CONTROL OF THE FORM-FITTED RAMP

Team 1629's ramp was designed to provide a maximum width for alliance robots. To achieve the desired width, the ramp was constructed on a set of drawer slides. In its stored position, the ramp was nested around the robot's elevator mast, with a hinged cutout of the platform floor folded up and resting on the base of the mast. The stored ramps fit snugly around the arm, but did not interfere with the arm's function.

When released, passive pneumatic pistons forced the ramp to unfold and the two platforms fell to the side of the robot. Once the platforms deployed and the ramp's weight rested on the floor, the driver moved the robot base forward. This motion allowed the robot base to move away from the ramp, and cleared the ramp from the elevator's tower. The resulting configuration prevented the arm from interfering with the ramp's function and provided the maximum area for mounting robots.

Propelled by gravity, this design eliminated the need for an additional motor to lower the ramp. The simplicity of the design and deployment contributed to the team's overall goals for reliable mechanisms and accurate control. A single motor was used to hold the ramps in their stored position. This motor was activated by two push buttons, with each button operated by each robot driver. The double-button protection prevented accidental release of the ramps since both drivers had to agree to release the ramps. The redundant buttons produced a fail-safe release process that prevented inadvertent deployment and avoided penalties.

COUNTDOWN FOR PERFORMANCE

Team 1629's robot included four types of sensors to monitor three robot functions and operate under two forms of robot control. These systems combined to achieve the team's single overarching priority: simplicity. High levels of performance were possible with simple mechanical systems that were easily monitored with standard sensors. Since each system was operated independently, the robot's control system was relatively simple. The simplicity resulted from creative ingenuity. Team 1629's focus afforded the team an opportunity to perfect each component, sensor and control routine, and produce a highly competitive robotic system.

▶ The arm's angle is measured with a rotary potentiometer to provide feedback and command the arm to the optimum position for placing a tube on the rack.

▶ To include a ramp system on the robot, the ramp had to be designed to accommodate the arm's mast. A small section of the platform is hinged to enable the ramp to be stored around the mast, with this section falling into place when the ramp deployed.

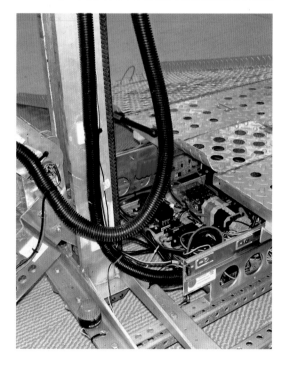

▶ The ramp system is mounted on a set of drawer slides connected to the robot base. The slides allow the ramp to surround the arm when stored, but is clear of the mast when deployed.

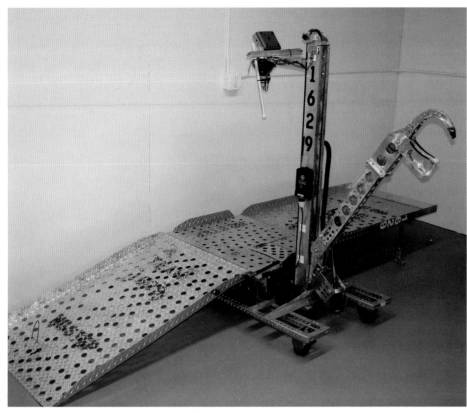

◀ Once the ramps unfold from their stored position, the robot base pulls away from the ramp platform to create an unobstructed path for alliance robots. In this position, the elevator mast is a safe distance from the climbing robots.

▼ Simplicity in design, operation, and control guided FIRST Team 1629 in its successful pursuit to create an award winning design.

Keeping Everything Under Control

▼ Tightly compacted in its starting configuration, Team 1731's robot folds its arm and ramp inside the robot to meet the volume constraints at the beginning of each match. The ramp is constructed from lightweight aircraft flooring and is strong enough to support two robots.

Heralding from Midland, Virginia, the Fresta Valley Robotics Team 1731 approached the control aspect of robot design with the same level of careful review practiced by NASA, their principal sponsor. With six points of articulation, their robot arm had a high degree of dexterity as a result of its accurate control. The team researched control methodology and made the best use of sensors and feedback.

Team 1731 designed a robot that was capable of both placing tubes and serving as a platform for their alliance partners at the end of the match. This combination of tasks required the team to efficiently pack the multiple functions into the allowed design envelope. The resulting configuration was one where the ramps were stored on the sides of the robot and unfolded to create a robot parking lot. The arm, which was long enough to grab tubes off the floor and score on all three levels, was similarly packed inside the robot and unfolded when needed.

The human body served as the platform for naming and understanding each component of the robot arm. Each of the six joints was referenced to its human counterpart and each required separate means of monitoring and control. The entire arm rotated at the base of the robot, with this function named the robot's waist. By pivoting at the waist, the arm rotated to retrieve and place tubes independent of the robot's orientation. Rising up from the robot platform, the next joint was the shoulder.

This joint was pneumatically powered and controlled with an innovative combination of computer-controlled valves. The system's elbow was powered with an electric motor using a large diameter cam at the joint to provide the required output torque. Being chain-driven, the motor was located halfway down the arm to minimize movement between the motor and the shoulder joint.

The final three joints were associated with the wrist and claw. The power source to rotate the wrist along the forearm's axis, referred to as supination, was similarly located near the previous joint to minimize the load on the elbow joint. The wrist's capability for extension and flexion—increasing and decreasing the angle between bones—was powered with a motor located at that joint. A pneumatic rotary actuator was used to open and close the robot hand.

Team 1731 demonstrated its control prowess in three distinct ways. They created a unique pneumatic control system to vary the shoulder position, fashioned a miniature arm to control the robot arm, and applied measurement and control theory to accurately position the robot arm.

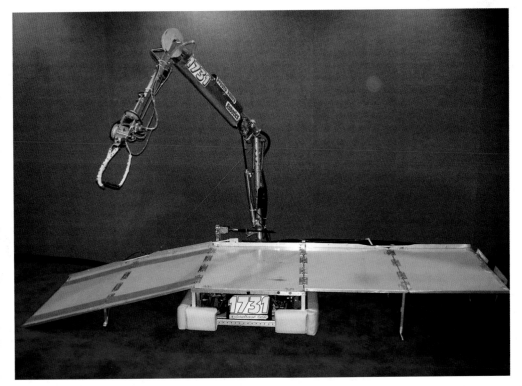

▲ Autodesk Inventor is used as a design tool to ensure system compatibility and performance. The tube manipulator system is capable of rotating around its base, thereby allowing the hand to grab and place tubes independent of the robot base's orientation.

◀ The wrist at the end of the arm is able to twist and rotate. A pneumatic rotary valve allows the hand to open and close as a means to grab and release the tubes.

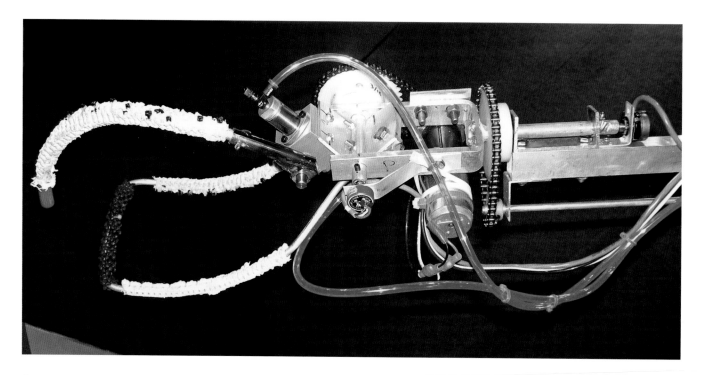

▲ A long driveshaft connects the motor and sprocket that rotate the wrist along the forearm's axis, with a separate motor for the extension and flexion functions. In each case, the motor speed is reduced and the output torque increased using a chain and sprocket system. The green tubes provide high-pressure air to the pneumatic rotary actuator that opens the hand.

UNIQUE PNEUMATIC CONTROL

The simplest use of a pneumatic cylinder involves injecting high-pressure air into one side of the cylinder. As the cylinder expands, the air from the other side of the cylinder is released into the atmosphere. An electronically controlled valve manages the system, alternating between powering and exhausting each side of the piston to either fully extend or fully retract the cylinder. The FIRST Kit of Parts includes a number of these two-position solenoid valves capable of positioning the cylinder in one of these two positions.

Since the shoulder was pneumatically powered, a full range of position control was needed. Getting this joint to work properly was the team's biggest challenge. Two types of solenoid valves were supplied in the FIRST Kit. Single-acting solenoid valves functioned as an on/off switch to supply high-pressure air to one side of a cylinder when in the on position, and to release the cylinder's captured air in the off position. Double-acting solenoids routed high-pressure air to either side of the piston while allowing the other end of the piston to exhaust. While a single-acting piston could extend a piston, a double-acting solenoid could repeatedly extend and retract a piston.

Team 1731's design combined both valves to create a pneumatic system that positioned the piston at any location along its stroke length. High-pressure air was directed by a double-acting solenoid into one end of a 1.5" (3.8 cm)-diameter bore Bimba cylinder. The exhaust port for this chamber of the solenoid was connected to the working pressure port of a second solenoid valve that served as a clutch. The first valve controlled the direction of the cylinder's movement, and alternated between supplying high-pressure air to either end of the cylinder. To control the cylinder's position, the clutch valve would be opened to bleed air from the exhaust end of the cylinder.

With the clutch closed, the cylinder would be pressurized at both ends, thereby locking the piston in place at any location along its stroke. To ensure that the forces on each end of the cylinder head did not generate pressures that exceeded safety limits, a pressure regulator was located between the two solenoid valves.

Initial testing of the pneumatic control system yielded stability problems as the arm bounced between the desired positions. The system hunted for its correct location and would overshoot its commanded location, only to again overshoot its location when attempting an approach from the opposite direction. To correct this deficiency, flow control valves located on the piston were adjusted to reduce the rate at which air entered the

Team 1731 "Challenger" Pneumatic Circuit

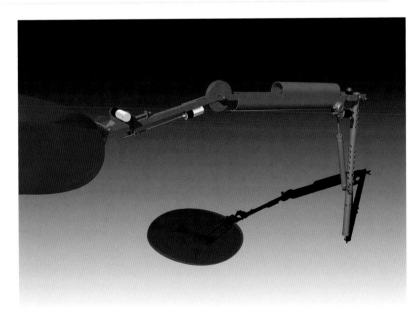

Thomas Air Compressor

Filter

One-way Restrictor

One-way Restrictor

Norgren Relief Valve

Clippard Air Receivers

Norgren Reglator

¾" Bimba Cylinder for Ramp 1

¾" Bimba Cylinder for Ramp 2

¾" Bimba Cylinder for Ramp 3

¾" Bimba Cylinder for Ramp 4

Pressure Guage

Nason Pressure Switch

Parker System Pressure Vent Valve

SMC SY3240 Single Solenoid Valve

SMC SY3240 Double Solenoid Valve

Norgren Reglator

Festo Valve as "Clutch"

Closed Path

Air Bleed

Festo Valve

SMC Flow Control

1 ½" Bimba Cylinder for Shoulder

Rotary Actuator for Hand (Grip)

▲ A schematic illustrates how all components of the pneumatic system are connected. Position control of the shoulder cylinder is achieved by connecting the exhaust port of a double-acting solenoid to a second solenoid that regulates the pressure at the rear of the piston.

▶ When the piston is fully retracted, the arm folds onto itself, and when fully extended, the shoulder is nearly vertical. Clever pneumatic control allows the shoulder to be positioned at any location between these two extremes.

cylinder. Also, the control system's dead band, the range near the targeted position where no control actions were commanded, was increased. With this change to the control system's software and the reduction in the system's response speed, the hunting phenomenon was eliminated and a full range of position control was achieved.

CONTROL WITH THE MINIATURE ARM

It was anticipated that control of the six independent joints of the robot arm would challenge the robot operators. Each joint was powered by a separate motor or piston, and fluid motion required that the joints be operated simultaneously. Using joysticks and switches for positioning the arm and operating the hand would require a significant amount of training, a high degree of coordinating actions, and the constant mental conversion to map stick positions to arm locations.

An intuitive method of arm control was achieved by constructing a ¼-scale model of the arm and using this model as a controller. Five of the six joints on the model and robot arm were monitored with a potentiometer, and a control algorithm was written to have the robot potentiometer values constantly match those of the model. A switch was used as the trigger for the hand. With this system, the arm operator did not have to monitor and affect each of the six joints independently. Rather, the controller simultaneously commanded each joint's power source based on two parameters established by the operator: the arm location and the state of the hand.

Rotational potentiometers on the model's waist, shoulder, elbow, and wrist measured the current location of each joint on the miniature arm. These signals were directly wired from the analog to digital conversion inputs on the robot controller. The robot controller used these measurements to determine the power that needed to be supplied to each joint. Potentiometers on the actual robot's joints measured the orientation of each joint, with the controller adjusting the power to each joint's motor or pneumatic system to bring the two potentiometer readings in agreement.

A limit switch mounted on the end of the miniature arm controlled the state of the robot's hand.

If the switch was in its default position, the hand was open, and if the switch was closed, the robot's hand closed. The integration of all functions on the miniature arm allowed the operator to completely control the arm using just one hand.

The miniature arm greatly improved the controllability of the robot arm. Being a smaller system with less mass and inertia, the miniature arm could be moved much faster than the real robot arm could follow. To compensate for this lag in response time, the robot operator was trained to maneuver the model at a pace that the real robot could respond to. This small amount of training was all that was needed to simultaneously manipulate each of the six arm functions.

CONTROL ALGORITHM DETAILS

Control consists of a combination of measurements and feedback. The relationship between these two functions is illustrated by a review of the control algorithm details to position the elbow joint. This example illustrates how each component is part of an integrated system for monitoring and positioning.

The orientation of the robot elbow was established by moving the miniature arm to a desired position. This orientation was measured by the rotational potentiometer on the model arm's elbow. While the potentiometer produced a voltage level between 0 V and 5 V at its two extreme positions, physical limits prevented full rotation of the elbow. As such, the potentiometer output was limited to 0.92 V and 2.25 V. This voltage was input into the operator interface where an 8-bit analog-to-digital converter produced a digital number between 0 and 255 that corresponded a voltage between 0 V and 5 V. Given the limited range of motion of the elbow, the actual digital value ranged between the values of 47 and 110. This number, which represented the angular position of the elbow, was transmitted to the robot controller as the computer code variable "p3_x."

The commanded value was compared to the actual position of the elbow by the robot controller. A rotational potentiometer on the elbow joint measured the joint's rotational angle. Because the robot controller had a 10-bit analog-to-digital converter it was capable of measuring 1,024 distinct

▶ A ¼-scale model of the arm serves as the controller to position the robot's arm. Potentiometers at each of the five articulating joints, as well as a limit switch for the hand, describe the desired location of the arm.

▶ The robot mimics the actions prescribed by the miniature arm controller. Potentiometers on each joint measure the relative angle at each location, with power supplied to the actuators to match the model and robot parameters.

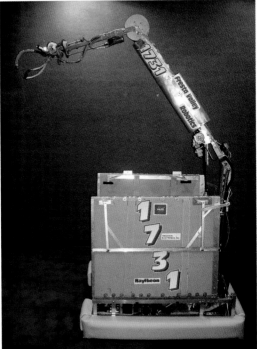

▲ Two potentiometers at the wrist joints on the control arm monitor the wrist's desired orientation. A limit switch at the end of the wrist is the trigger to open and close the robot's hand.

▼ CAD drawings and a block diagram describe the control circuit to position the elbow. The desired elbow location is measured with a potentiometer on the scale model, with a second potentiometer mounted on the robot elbow joint to measure the actual joint's angle of rotation.

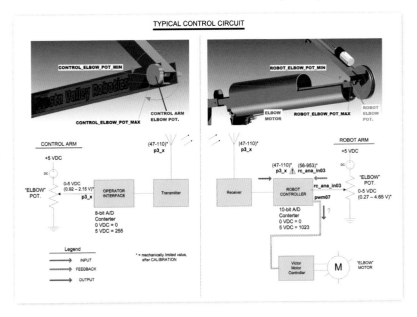

TYPICAL CONTROL CIRCUIT

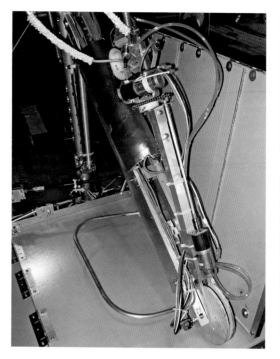

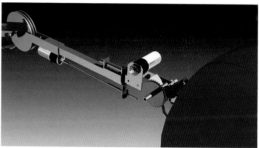

```
ACTUAL CODE SNIPPET

/* compute the ELBOW control */

/* READ the CONTROL ARM elbow potentiometer */
control_elbow_pot_value = p3_x;

/* READ the ACTUAL ROBOT elbow potentiometer */
actual_robot_elbow_pot_value = Get_Analog_Value(rc_ana_in03);

/* SCALE to robot elbow pot desired reading using INTERPOLATION */
desired_robot_elbow_pot_value = ( ( (control_elbow_pot_value     - CONTROL_ELBOW_POT_MIN) /
                                    (CONTROL_ELBOW_POT_MAX  - CONTROL_ELBOW_POT_MIN) ) *
                                  (ROBOT_ELBOW_POT_MAX    - ROBOT_ELBOW_POT_MIN)    ) +
                                    ROBOT_ELBOW_POT_MIN;

/* CALCULATE elbow MOTOR CONTROL value */
pwm07 = Limit_Mix(2000 + (int)( 0.75*(desired_robot_elbow_pot_value - actual_robot_elbow_pot_value) + 127));

/* INCORPORATE "DEAD ZONE" */
if(pwm07 > 120 && pwm07 < 134){
    pwm07 = 127;
}
```

◄ A snippet from the team's computer code illustrates the commands that map desired positions to motor control actions. Each of the robot's five control functions employs the same technique for intuitive operation.

▲ Separate motors power each rotation axis on the wrist. An extended shaft is used to move one motor closer to the elbow to reduce the momentum created by this motor. Potentiometers on each rotating axis measure the system's orientation.

levels. As such, the elbow's position was measured with a higher degree of resolution. The measured signal, referred to in computer code as the variable "rc_ana_in03," ranged between the digital values of 56 and 953.

The robot controller was programmed to scale the commanded position and compare that value to the actual position. If the two signals did not match, the robot controller supplied the elbow's motor controller with a signal to power the motor. Still working in a digital state, a motor controller command of "0" operated the motor at full power in reverse, and a value of "255" commanded the motor controller to supply full power in the forward direction. Values between 0 and 255 were proportional to motor speed, with a value of "127" commanded for the motor to stop.

The process of measuring the desired input, comparing that measurement to the actual elbow position, and commanding the elbow motor to correct for any discrepancies was constantly performed by the robot control system. In addition to conducting the operations for this one joint, four other joints were similarly monitored and positioned. Simultaneous with these functions, the status of the hand was measured and directed. As

if that wasn't enough, the robot control system also monitored and determined the power supplied to each of the robot's four drive motors. A well-written feedback control algorithm was essential to keep all of these functions under control.

USING CREATIVITY TO GAIN CONTROL

Team 1731's ingenuity and creativity were critical factors that enhanced the team's ability to accurately and easily control their robot. When faced with provided components that only allowed a piston to be placed in two positions, the team combined components to regulate the pressure supplied to each end of the piston. This creative combination produced a positioning system that could hold the elbow in any orientation.

To efficiently control the arm, the team designed a scale model of the arm that transformed the arm operation into intuitive control. To accurately position each joint in the desired location, a control algorithm was written to measure the commanded and resulting joint positions, with power supplied to correct any discrepancies between the commanded and resulting signals. All in all, Team 1731's innovating ability produced a robot that was always under expert control.

▲ A coordinated system of control gave Team 1731 the ability to score at any level. The tube could be twisted and rotated by the hand to find the best orientation when being placed on the rack.

Xerox Creativity Award

The Xerox Creativity Award celebrates creativity in design, use of a component, or strategy of play. While one team may be recognized for a single creative component, another team may have won this award for its creative method of scoring points. The award promotes ingenuity and originality. The award showcases unique designs that offer a competitive advantage. Teams that win this award progress from a creative idea to a creative product, and that journey is often not an easy one.

Xerox is a founding sponsor of FIRST. The company provides generous team and regional event support as part of its long-standing commitment to education and as an investment in the future diversity of its workforce. A former chairman and CEO of Xerox is also chairman emeritus of the FIRST board of directors, and Xerox's current president now serves on FIRST's board.

Kicking It up a Notch

The time limitation of each FIRST match is an important determining factor when analyzing strategy. With only two minutes and fifteen seconds per match, the ability to score game pieces sometimes relied heavily on the speed of a robot to collect and maneuver these pieces. In addition, the octagonal central rack provided 360 degrees of access, opening up scoring potential. Teams would not necessarily need to push each other to gain access to a spider leg. A desire for robot maneuverability and speed guided Team Krunch, from Clearwater, Florida, from early on in the design process.

IDENTIFYING THE CHALLENGE FOR A CLARIFIED APPROACH

The process of manipulating the game pieces was broken into three main actions: picking up, holding onto, and releasing a piece to score. To dependably perform these actions, the robot needed to have a sturdy and repeatable grip on the pieces, regardless of the approach angle or field of view of the human operator. To build a mechanism that could pick up pieces only during certain ideal situations is useless—one can never rely on ideal situations once the match action begins.

The final design for game piece manipulation included a kicker, which kicked the pieces from a horizontal position on the playing field floor to a vertical orientation. A claw grasped the game pieces from this position to place them on the rack for scoring.

▲ The claw of Team 79's robot places a keeper onto the middle level of the rack. Speed, reliability, and repeatability are essential to the 2007 challenge.

The movement of the game pieces from horizontal to vertical positions made the robot very visible to the robot operators. From across the playing field, students could easily see that a game piece was in position to be grabbed by the claw. Game pieces were picked up on the fly, without pausing for defending robots to interfere.

CONCEPTION OF THE KICKER

The kicker comprised two main parts: a primary and secondary arm. The primary arm supported a constantly rotating wheel, which made contact with the surface of a game piece to roll it up toward the robot and into an open claw. Initially, this primary arm was the only component of the kicker. The pieces would contact the wheel and occasionally slide rather than roll due to the low friction with the carpet, causing them to become pinched under the wheel-support arm. The secondary arm was added to remedy the unpredictable game piece motion. A series of free-spinning PVC rollers curved along this arm. Now, when the robot approached a piece, it became trapped between the spinning wheel and rollers, which helped guide the piece throughout its rotation. This rotation not only repositioned the game pieces, but also automatically centered and aligned the pieces for the claw. This provided the repeatability the team was looking for in a grabbing device while still maintaining the speed of the process.

These two arms were controlled by a single pneumatic cylinder, which extended the kicker past the front bumper to collect a game piece, and then retracted it into the robot frame once the piece was in the grip of the claw to prevent interference.

Through experimentation and prototyping, the team was able to decide what kind of wheel to use for rotating the game pieces. The best combination of wheel size and tread traction would ensure an accurate and controlled motion. The ideal height of the wheel to ensure solid contact between it and the game piece was found using Autodesk Inventor. A prototype tested the reliability of the system given possible variances in game piece inflation and carpet friction.

Calculations determined the rotation of the game piece from horizontal to vertical alignment to be slightly less than 135 degrees. The average diameter of a piece required approximately 11" (27.9 cm) of constant wheel contact to put the piece into its new orientation. A 6" (15.2 cm)-diameter wheel would provide this contact in approximately one full rotation. Initial designs had the wheel supported by Lexan plates, which yielded to cracking due to rough defensive play on the field, and prompted the team to replace these structural members with more robust aluminum plates.

▲ The method of rolling the game pieces with the kicker assembly positioned each piece in the same orientation every time. This repeatability greatly benefited scoring speed.

▼ When the kicker is fully extended by the pneumatic cylinder, it eagerly accepts game pieces. The free-spinning PVC rollers of the secondary arm prevent game pieces from being pinched under the primary arm.

▶ The retracted kicker assembly pulls back into the robot frame, freeing up space for the claw and arm to transfer the game piece to the rack.

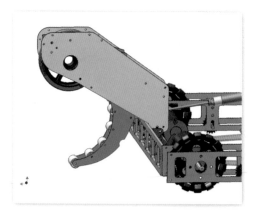

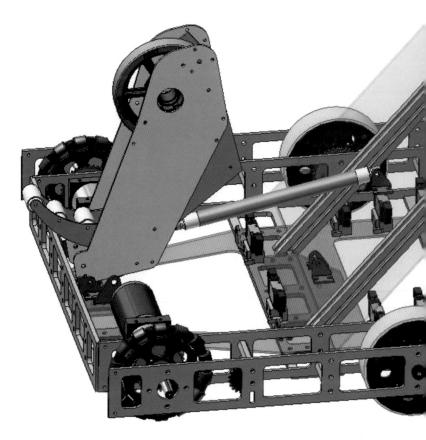

▶ Autodesk Inventor provides detailed system models to determine the ideal geometry needed for the kicker assembly.

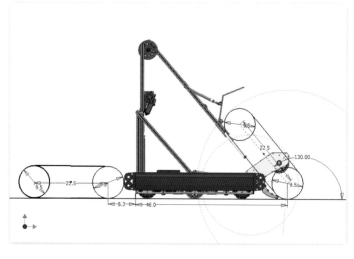

TRANSFERRING GAME PIECES TO SCORE

Arm and claw mechanisms were constructed to navigate the game pieces from the kicker to a spider leg for scoring. Early suggestions of a claw were prototyped to test the ability of collecting game pieces from the floor, before the kicker idea was introduced. The claw provided excellent grip, but was not efficient at picking up the pieces. With the kicker implemented, the initial claw designs were further developed to grab a game piece from a predictable position and deposit it for scoring.

The first claw utilized a pneumatically controlled dual-action finger and thumb to grip the game pieces. When this claw was used on a field in competition, Team 79 found that the grasp was not strong enough, and opposing teams could disengage the game pieces. Two more fingers were added to the claw, which provided sufficient support to surmount the dilemma.

The claw was attached to the end of an arm, which was supported by a tower mounted to the robot chassis. The tower was designed to lift the arm to reach the highest level of the rack, and was mobilized by the already-present pneumatic system implemented for the claw. Autodesk Inventor was again used to model the spider leg heights, game piece position, and other parameters. From these models, the correct arm length and overall design was refined. Low-speed window motors with worm gears provided in the Kit of Parts drove the arm's rotation, and resisted counterrotation.

The large momentum of the swinging arm minimized stability and control. The system was stabilized with a gas spring coupled with a cable and pulley system around the axis of arm rotation. The spring provided a counterforce against the torque of the arm generated by gravity and allowed the arm to maintain its swift delivery of game pieces.

◀ The robot approaches a game piece with the kicker extended. The piece engages with the spinning wheel on the primary arm and rollers on the secondary arm, to begin its rotation to a vertical orientation.

▼ A game piece is flipped up to the claw, which grasps the piece with three pneumatically controlled fingers and a thumb. The kicker is retracted back through the robot frame to avoid interference with the claw.

▶ The arm raises the game piece over the robot to be placed on a spider leg for scoring. The arm is attached to a pneumatically liftable tower to reach the highest rack levels.

▼ Autodesk Inventor drawings of the rack and robot help the design team with calculating the necessary arm and tower dimensions to support the potential of scoring on any rack level.

◀ The robot is positioned with the backside facing out, to allow the camera to face the rack and maximize time to track the target light. The strategy of starting the autonomous round with a keeper, and directly scoring behind its back on a spider leg, became known as a reverse jam.

▼ A creative method of picking up game pieces and depositing them on the rack defined and empowered Team Krunch's machine.

REINFORCING SUCCESS

A CMUCam2 was mounted on the back of the robot to detect the target light, where it was also protected from damage. This position enabled the sensor to stay locked on the target for a longer amount of time. When a game piece was secured in the claw, the arm rotated over the top of the robot, and scored from the camera side. In autonomous mode, the robot did not have to turn around before approaching the rack, resulting in a higher chance of successful game piece placement.

To protect against defensive maneuvers by opponents, especially when placing game pieces for scoring, a braking system was added. The onboard pneumatic system was again used to activate cylinder-controlled brake pads at each corner of the robot.

Team 79 anticipated the fast field action required of the 2007 challenge. Their visualization enabled them to maintain focus on their design and apply their creativity to design a kicker to streamline and regulate game piece manipulation. Their imagination and craftsmanship was made evident when it came down to "Krunch-Time"!

Royal Assault: Patents Pending

Creativity is defined as the ability to transcend traditional ideas. The creation of new solutions to a common challenge is one of the driving forces behind FIRST robot design. This progress is rewarding, especially for Team 357, from Drexel Hill, Pennsylvania. Their unique approach to, and exploration of, mechanical solutions is driven, not hindered, by encountering physical restrictions. The product of their inventiveness was a standout machine that incorporated novel devices to execute their strategy.

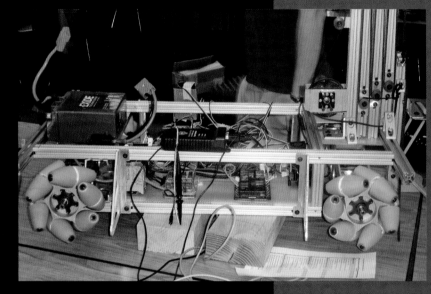

▲ The six sets of rollers on each wheel of the omnidirectional Jester Drive provide the traction and maneuverability needed to navigate the playing field.

REVISING AND IMPROVING THE JESTER DRIVE

The Royal Assault team anticipated the necessity of maneuverability for success in the 2007 challenge. In past seasons, the use of a Mecanum wheel, developed by Airtrax, Inc., has been the signature of the team's Jester Drive. This wheel design utilizes multiple rollers mounted at 45-degree angles around the edge of a wheel. Each dual-layer roller comprised a solid urethane inner core with a softer outer layer for increased traction. These rollers are mounted on separate shafts and divided to each half so they can rotate independently. The resulting design allows omnidirectional movement, giving the appearance of the robot gliding across the playing field.

The Jester Drive was used in the 2006 season with great success, however the tread layer on the rollers separated from the core component, and the tips became deformed after prolonged use. Modifications of the preexisting design could improve the traction, stability, and overall performance of the drive.

A review of the problem indicated that the bond between the layers degregated over time due to the lack of durability of the chemical bonding agent used to join them. A physical bond was created between the two layers in the form of a series of slightly tapered parallel grooves. These grooves helped improve the adhesion by increasing the surface area on which to mount the tread layer.

An extended, smaller diameter roller axle on a 6'' (15.2 cm) wheel replaced the design of the 2006 robot. The new axle, along with a reduction of tread thickness, allowed a thicker roller core for added support.

The modifications to the rollers were drawn in Autodesk Inventor, and 3D models were printed in ABS plastic using stereo lithography. Silicone molds were fashioned, and from a dual molding process, the hard inner and soft outer urethane components were fabricated. The renovated rollers exhibited increased traction and showed no evidence of delamination after extended use.

INVENTING A NEW SOLUTION

To incorporate the ability to elevate game pieces to the separate rack levels, a series of telescoping 80/20 extruded aluminum pieces was fitted together. The team searched for lightweight methods to connect the pieces of aluminum, but found a solution only in the form of a separate section of 80/20 to support fixed external glide brackets. Rather than sacrifice this device for a lighter alternative, Team 357 instead applied their knowledge of molding to fabricate their own solution, the Jester Glide.

A mold was created that would capture the shape of the space between two channels. A self-lubricating material in this shape would support the two sliding members. The mold was made from two sections of 80/20 clamped together and separated by two pieces of 1'' × ⅛'' (2.5 cm × 3 mm) aluminum bar stock spacers on each side of the center channel. SmoothCast Roto, a fast-curing urethane plastic, was poured into the mold to create a rough model of the linear bearing. This model was fine-tuned to a specified tolerance for acceptable clearance between the aluminum and bearing and a refined silicone mold was created.

▲ The prototype mold is made from two pieces of 80/20 extruded aluminum clamped together. The space in the center is the desired shape of the self-lubricating linear bearing.

▶ Polyester fiber, urethane plastic, and molybdenum disulfide powder are combined and poured into a refined silicone mold. When set, the finished Jester Glide is removed from the mold.

The final bearings were made from a mixture of SmoothCast Roto, molybdenum disulfide for lubrication, and ⅛'' six-denier polyester fiber for strength. Different proportions of these materials were experimented with to find the desired friction, strength, and durability. After molding, the ends of each piece were sanded and installed with a setscrew. To prevent the glide from slipping out of the channels, a hard stop was created at the end of the channel by removing one of the channel profiles from a piece of Jester Glide.

The new plastic linear bearing performed just as well as the commercially offered linear glide; at less then 90 percent of the weight of the linear system, a total of 3 lb. was saved. Students searched for a similar self-lubricating linear bearing for 80/20 extruded aluminum, but found none offered. This motivated them to apply for a provisional patent, submitting drawings made during the robot design as a part of the application.

NEW APPROACH TO AN OLD PROBLEM

Every year, a common task faced by every team is to mount the battery in an accessible, secure, and safe place on the robot, without adding extra weight. Once again, Team 357 seized the opportunity to demonstrate their mastery of molding. The first mold was designed to fit the backside of the 2007 competition battery. This mount was small and lightweight, just as designed, but did not fit batteries from previous competition years. The mounting plate was further adapted to incorporate battery configurations from previous competition years, until a final design securely cradled the backside of all different forms. After this, the team designed a mounting plate that supported the front of the 2007 battery, which was again modified to accept the different battery configurations. The team applied for a second patent for the finished universal battery-mounting device, which accepted all styles of FIRST regulation 12 V batteries.

▲ The lift sections are assembled with Jester Glide. The small piece on the outer edge acts as a hard stop to prevent the Jester Glide from slipping out of the channel during operation.

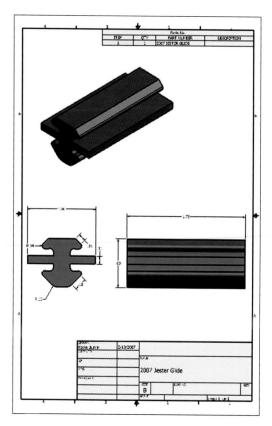

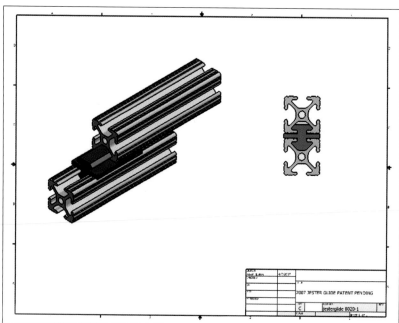

▲ Drawings of Jester Glide depict how it fits between the channels of the 80/20 aluminum. Such drawings were submitted as a part of the patent application.

A SUBSTITUTION FOR TRICKY RAMPS

Primary strategy planning determined that the critical make-it or break-it point of the challenge would be the potential bonus points at the end of each round. The ability to lift alliance partners could be the deciding factor of the winning alliance. Rather than rely on the ability of alliance partners to be able to maneuver and climb a steep ramp, Team 357 wanted a reliable, predictable, and repeatable method of scoring bonus points.

The solution to raising the alliance came in the form of horizontally elevating platforms. These platforms, one on each side of the machine, were initially positioned at ground level when deployed. The alliance teams needed to only overcome the thickness of the platform to drive onto it and park.

The low profile, transparent platform was constructed of lightweight yet strong Alumalite, a twin-wall polycarbonate. The scissors were mounted above the platform, one on each side, in a starting position that was calculated to be 4'' to 5'' (10.2 to 12.7 cm) open. Fisher-Price and BaneBots motors were mounted on a piece of aluminum angle at the base of the platform, one powering the scissor on each side. The motors turned a lead screw that was attached to one leg of the scissor lift, with the opposing leg riding in a horizontal slot. As the motors turned, the platform was uniformly elevated to over 12'' (30.5 cm). This lift required only the level positioning of alliance partners. The need to climb or navigate narrow ramps was eliminated, thus fulfilling the reliability required by team design.

Rather than succumb to the limitations encountered in the 2007 season, Team 357 used their imagination and intellect to overcome obstacles. Improving upon their already noteworthy drivetrain, inventing linear bearings and battery mounts, and a facilitating lift system demonstrated that Royal Assault is prepared and eager to "Dare mighty things."

▲ The top 80/20 piece of aluminum is connected using the pre-existing linear glide with brackets. The lower piece is attached using Jester Glide, the thin black material between the two channels.

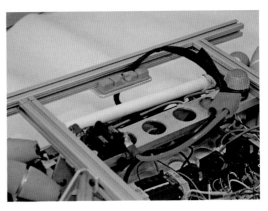

▶ The universal battery mount plate is specifically designed to securely hold the FIRST regulation 12 V batteries, from both the 2007 season and previous years.

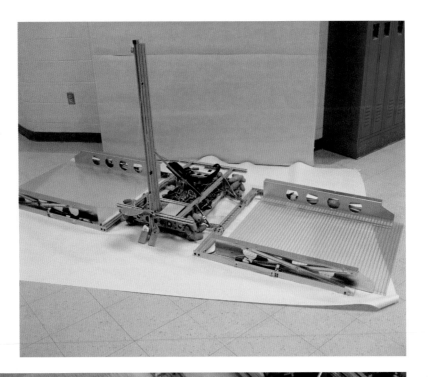

A low profile, transparent platform is easy for alliance robots to mount. The lightweight, yet strong polycarbonate material provides sufficient support to lift the potential 120 lb. on each side.

To activate the scissor lift, a motor drives a lead screw that is connected to one leg of the lift. When driven, the platform is raised sufficiently to qualify for bonus points.

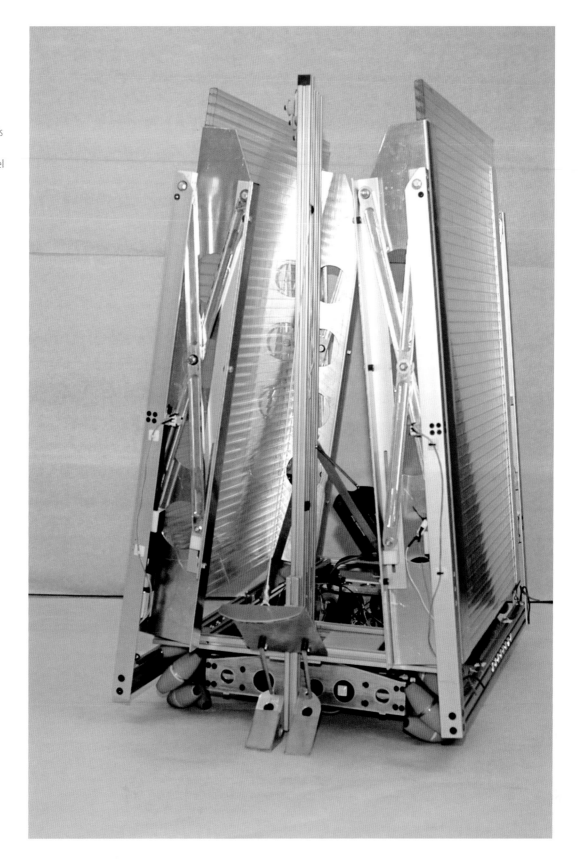

▶ With platforms folded for the start of a match, this machine is equipped with innovative solutions to excel at the 2007 challenge.

Reducing Complexity for Construction

Robot repair and adjustment in the limited time between consecutive matches can be hectic. Hardware and spare parts scattered around a pit create an inefficient working atmosphere. Team 1100, from Northborough, Massachusetts, has mastered the art of component simplification. They use their creativity to approach a challenge in search of the most effective and uncomplicated solution. A three-step approach is employed to maintain clear focus on the task at hand.

SKETCH, PROTOTYPE, BUILD

As soon as the 2007 challenge was announced, the ideas began to flow. Analysis of the game inspired the team to choose a vertical lift with a wrist-mounted grabbing mechanism. It was decided that a linear motion lift would provide the fastest and most effective way to raise and lower the grabbing mechanism.

Algonquin Robotics is a small team, with limited resources in the areas of machining and computer-aided design programs. In this sense, the design process is simplified. While there are an infinite number of ideas for mechanisms, only those that are feasible are considered. This narrows down the design options and forces the team to focus on accomplishing its specified goal. The process is referred to as the "sketch, prototype, build" method. A rapid succession of these three steps, often repeated for all components, gradually produces a final design. Rather than initially create detailed designs of the robot components, they are instead added and modified as the robot evolves.

▲ A rapid method of sketch, prototype, and build is the process employed by Team 1100 to narrow design options down to a final robot concept.

The design process began as a series of sketches to define the robot and its functions. The team then reviewed the sketches to identify possible improvments. The reviews generated of a list of suggestions, and often LEGO models were created to communicate these ideas in a three-dimensional form. A physical model of the system could be visually manipulated as it developed and evolved. When a design was selected, refined sketches were made using Microsoft Viseo, a diagramming software that utilizes vector graphics, to provide further detail and dimension.

The next step in Team 1100's process was prototyping. Full-size proof-of-concept models were often built to verify that a certain mechanism was feasible. These models could then be altered to perform in the same way that the final machine would. The first design presented for the linear motion lift incorporated a tubular material and linear bearings mounted in blocks. Height would be achieved as the blocks slid up the tubing. This idea was prototyped and functioned, but would not fulfill the requirements of speed, efficiency, and ease of build set by the team. A new vertical lift design was investigated to accomplish the task of raising a gripping mechanism.

SYMMETRY SIMPLIFIES REPAIR

The new design incorporated structural members made from 80/20 extruded aluminum. Three mobile modular stages extend upward from a fixed section when activated to provide sufficient reach for scoring on the top spider legs. Each stage has its own self-contained chain, enclosed within the 80/20 aluminum with accompanying pulleys. The system could have been designed with one continuous chain running through all three sections, but the complexity of building and maintaining such a system was impractical.

The modular system enabled easy replacement of a stage if it needed repair. The lower stage was mounted to motors and a fixed section attached to the robot chassis. The middle and upper stages of both the left and right assemblies were identical. If either were to become damaged during play, a single spare part would be adaptable to either position.

The stages were designed to require little effort for assembly and disassembly. Each stage was attached with only four screws and all fittings on the three stages were interchangeable. With such little hardware incorporated in the system, repair was not a laborious task.

▲ A LEGO prototype helps present ideas in a three-dimensional form. Team members can better visualize the possible strengths and weaknesses of different concepts.

A three-member modular lift system relies on 80/20 extruded aluminum structural members to encase the inner drive chains and pulleys. Lift teeth plates can be seen at the upper ends of the members.

The closed chain loop enables the three lift sections to move simultaneously. The lift can extend from a closed position to maximum extension in fewer than two seconds.

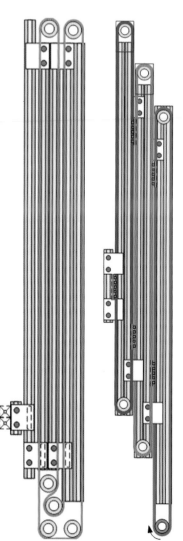

A single long chain running through all three sections can control the motion of the lift. The team decides that separate chains for each stage would be a less complex and more efficient design.

UNIFICATION OF THE MOBILE LIFT STAGES

The vertical lift was designed with a gap of approximately ⅛'' (3 mm) between each stage. To function properly, it was necessary to attach the walls of one lift section to the chain running inside the subsequent section. Each stage was powered by the previous stage. To connect the adjacent lift stages, a lift teeth plate was fabricated. Six holes were drilled down the centerline of a ⅛'' (3 mm) thick aluminum bar. Six 4/40 screws were inserted into these holes, to line up and mate with gaps in the chain. While this device worked for attaching two sections of the lift, it was not strong enough to support the weight of all three mobile sections. Excessive stress bent the screws and the lift components fell to the floor. This device could not withstand the forces imparted on it, and it was inadequate to transfer power from the lower stages to upper stages.

FINITE ELEMENT ANALYSIS

The challenge of connected abutting lift stages had to be reevaluated. The screws were not strong enough to support the weight of the lift, but there were no other solutions without custom designing a device. Students began work on a special design that would satisfy the strength requirements. A solid metal block, representing an unrolled sprocket, was simple enough to build without the necessity of elaborate machining. Detailed design engineering was performed and a drawing of the piece was created using SolidWorks.

From the SolidWorks model, the part could be further studied with a SolidWorks factual design analysis tool, COSMOSWorks. This program enabled performance of finite element analysis on the new plate, and tests determined that the plate was strong enough to support more than a full robot's weight. This information was sufficient to begin construction on the multiple plates that would be needed for the system. A manual mill was used to fabricate these plates, with dimensions controlled by dials and gauges. The students, who carefully machined the twelve plates, controlled the precision needed to ensure the chain meshed with the new plate.

Finite element analysis of the plate design enabled testing the system with theoretical conditions. The safety of a system is defined as the ratio of the breaking stress and the applied stress, or the amount of load that a mechanism can handle without failure. A dimensionless factor of 1 denotes that the applied stress is equal to the allowable stress before breaking; anything less than 1 implies failure. The safety factor of the plate, at its lowest value, was 5.264, well above failure conditions. A static strain analysis located the points under the highest stress, at the base of each tooth. This software confirmed that the replacement plate would hold up to the rigors of competition.

For Team 1100, the tried-and-true method of sketch, build, and prototype again produced a successful machine. The straightforward lift design improved time management, both on the field and in the pit. Dedicated students applied their simplified approach to the invention of the lift teeth plate, which was effective owing to meticulous construction practices. The design of this machine is evidence that elaborate, complex components are not always a gauge of a successful robot.

▶ The original lift teeth plate failed when the inserted screws bent under the weight of the assembled lift system. A redesign produces a single plate with teeth that can withstand the applied forces.

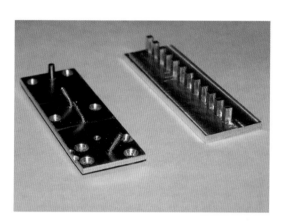

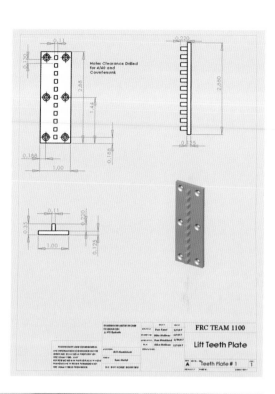

▶ A CAD drawing of the redesigned lift teeth plate clarifies the exact dimensions required for even meshing with the chain. This drawing enables performance testing using finite element analysis.

▲ Students manually mill the lift teeth plate, taking care to adhere to the size specifications. The completed plate meshes perfectly with the chain.

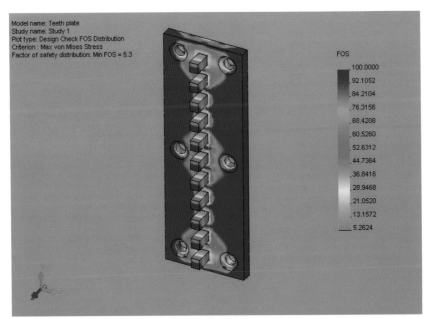

Model name: Teeth plate
Study name: Study 1
Plot type: Design Check FOS Distribution
Criterion : Max von Mises Stress
Factor of safety distribution: Min FOS = 5.3

FOS

100.0000
92.1052
84.2104
76.3156
68.4208
60.5260
52.6312
44.7364
36.8416
28.9468
21.0520
13.1572
5.2624

◄ Finite element analysis is used to find the factor of safety of the plate design. A value over 1 ensures that the component will not fail under the applied stresses of the lift system.

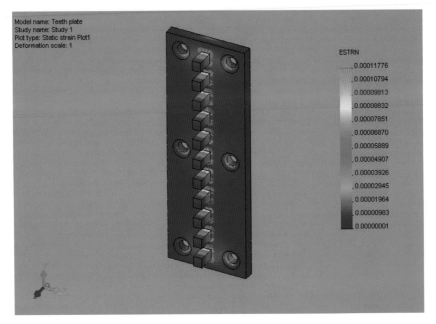

Model name: Teeth plate
Study name: Study 1
Plot type: Static strain Plot1
Deformation scale: 1

ESTRN

0.00011776
0.00010794
0.00009813
0.00008832
0.00007851
0.00006870
0.00005889
0.00004907
0.00003926
0.00002945
0.00001964
0.00000983
0.00000001

◄ The static strain distribution is plotted on the lift teeth plate using COSMOSWorks. The highest stresses occur at the base of the teeth.

▶ The completed lift system incorporates three modular sections, connected to each other with a custom-designed lift teeth plate. This design efficiently elevates game pieces to the desired spider leg height.

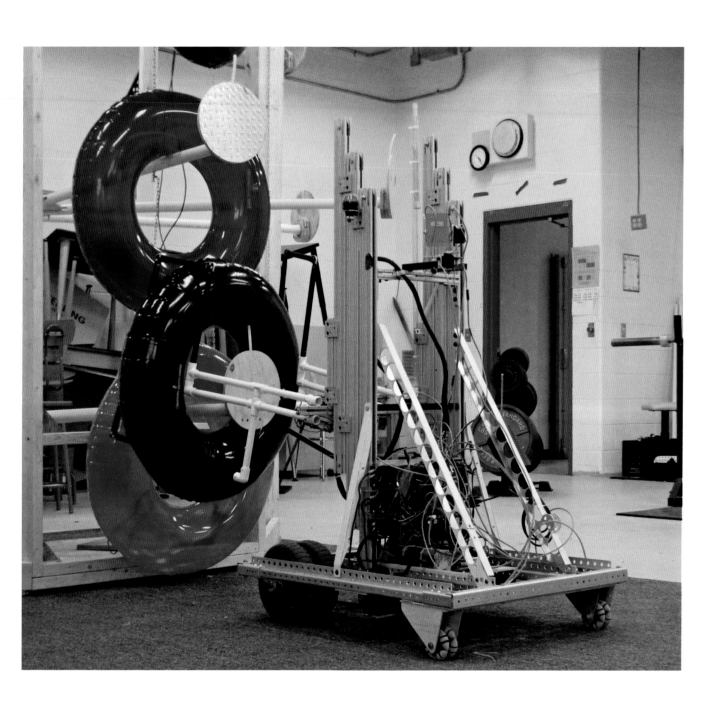

Accommodating an Alliance

The announcement of the 2007 game sparked an ambition among members of Team 1319 to build a robot that could take advantage of every method of acquiring points. Golden Strip Flash instantly immersed themselves in design proposals that could meet every challenge. Hailing from Mauldin, South Carolina, this enthusiastic team recognizes that teamwork strengthens any alliance. The robot not only played the field, but also was specifically designed to facilitate interaction with alliance partners. Strict adherence to allocation of component weights ensured that the final assembly of all elements did not deliver an excessively overweight robot. The weight goals set early on in the design process helped subgroups retain awareness of this limitation.

FINGERS, ELBOWS, SHOULDERS, AND ARMS

To realize the goal of participating in every scoring activity, the robot design included a device to unite game pieces with the spider legs. The team began designing an arm and gripping mechanism to manipulate the game pieces. The device needed to fit within the starting size limitations and weigh no more than the 15-lb. goal set early in the design process.

Prototypes of various gripper designs were fixed to an arm to determine the most efficient way to interact with the game pieces. Devices included two fingers to pick up a piece from the center hole, rollers to capture the pieces, and fingers to grab the piece from its outer edge. Performance of these prototypes was evaluated until one design was agreed upon. A four-finger claw featured a strong grip on each game piece and had manageable control for the drivers in the alliance station.

▲ A four-finger design grips game pieces securely. The bottom two fingers are slightly sloped to slide under a piece. An actuator closes the top fingers when a game piece is in place.

The fingers of the chosen design were made of PVC pipes with 0.5" (1.3 cm)-thick walls. Two sloped bottom fingers guided each piece into alignment with two upper, angled pipe fingers. An actuator mounted on the arm extends to close the upper fingers around a game piece. The four contact points of the grip prevented the game pieces from being knocked loose or dropped, quashing defensive maneuvers of opponents. The finger arrangement was able to pick up game pieces from the floor and alliance station wall and place or remove them from the rack.

The arm was constructed of aluminum tubing, with supports added for increased robustness. Two pivot points enabled the arm to reach from the floor to the top of the rack. A chain and a window motor coupled with a DeWalt gearbox rotated a shoulder joint to adjust the elevation of the arm. An actuator acted as the elbow joint, which extended and retracted the arm. The arm provided the gripper with the necessary mobility to hang game pieces on all three rack levels.

▶ An actuator controls the extension of the arm. A sensor prevents the arm from being driven into the vertical supports.

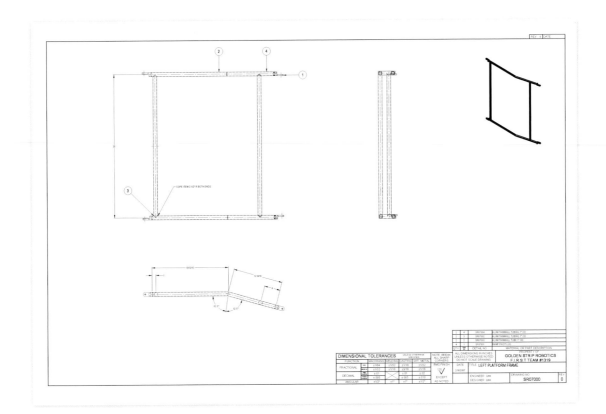

The ramp and platform are framed with aluminum tubing. CAD drawings help team members visualize and measure the necessary dimensions of the bend in the frame to create an incline angle of 15 degrees.

With the arm rotated upward and elbow extended, the gripping mechanism easily lifts game pieces to the highest level of the rack for scoring.

REDUCING THE ANGLE OF INCLINE

Although the original intent was to produce a robot that could perform in all scoring activities, the machine was quickly approaching the 120-lb. weight restriction, despite the appointed component weight distribution. With only 25 lb. available, the decision came down to adding one of two mechanisms. The potential strategies for scoring were examined. The robot, named Mick, could either be fitted with ramps or include a method to earn points in the autonomous period. The potential bonus points available for a ramp system were much more valuable, so the choice easy; the autonomous strategy was abandoned.

Ramp design started out with a single goal: develop an easy-to-mount device that can lift two alliance robots simultaneously. This would be accomplished with a ramp and platform combination. The first step in developing such an apparatus was to make a prototype robot chassis to better visualize integration. It was revealed that if a ramp were extended from a 12″ platform (the minimum height required for bonus points) to the floor, the resulting incline would be 22 degrees. This steep

angle could potentially hinder the ability for other robots to climb on the platform. Wanting to reduce this angle, the team now faced particular design requirements of a low incline and wide ramp, with a platform long enough to accommodate two full-size robots.

Computer modeling, in particular AutoCAD, enabled the visualization of components, clarifying the dimensions and layouts so the ramp system would fit requirements without interfering with other robot functions. Solid Edge, a three-dimensional parametric program, was then used to further refine the design and provide more detail to each part.

To decrease the angle of incline, the ramp was divided into two parts. One stage attached to the robot chassis, with a slight bend in one end of the frame. Extending from this bend was another stage that folded out. The combination of the two sections lengthened the overall ramp, resulting in an incline of only 15 degrees.

DEPLOYING THE RAMPS IN TWO STEPS

The first step to ramp deployment was the unfolding of the two ramp sections. An actuator was connected to the frame of both the first and second stages of the ramp. When the actuator was triggered, the extension began to unfold and raise the first ramp stage.

The second step to lower the ramp and platform was the release of the components from their vertical orientation. The ramp and platform were held in an upright position with a hold-release mechanism. A shoulder bolt on each side of the robot attached to the platform and ramp. A common actuator held these shoulder bolts with a cam at each end. When the actuator was retracted, the shoulder bolts held the ramp and platform in upright positions. Activation of the actuator released both bolts and the cams pushed the ramp and platform out to separate sides of the robot, where gravity took over to fully lower them. The release of the shoulder bolts enabled rapid deployment of both mechanisms.

Legs supported the ramp and platform at full extension to maintain a level elevation. These legs were initially folded up with the two stages of the ramp, but had to rotate 90 degrees to be positioned perpendicular to the floor after the ramp and platform were deployed. Adjustable rods attached to the base of the robot extended out to the legs. These rods pushed the legs of both the ramp and platform into position as they were lowered, to evenly support alliance robots if necessary.

A wide ramp and platform elevated a robot 13.5" (34.3 cm) above the playing field floor. The additional height accounts for potential sagging, which could cause the guest robot to dip below the 12" (30.5 cm) mark.

Aluminum tubing was used for the frame of the ramp and platform for its strength, light weight, and durability. Sheets of 11-millimeter Polygal, a unique, lightweight, and strong material that is often used as hurricane sheeting to protect windows, covered the frames. The surface is smooth, so a tread was added to facilitate the climb for other robots.

The decision to forgo the autonomous capability yielded a ramp and platform that could really "rack up the points." An enduring determination to maximize the performance of the robot roused the creative minds of Golden Strip Flash. The result was a machine that alleviated the burden teams encountered when climbing a steep and narrow ramp to score bonus points. A wide ramp with a reduced angle of incline and large platform enabled two robots to easily surmount the platform to settle well above the requisite height. This dynamic machine accommodated alliance partners, exemplifying the profitability of teamwork.

▲ The ramp is folded up at the start of each match. An actuator is connected to both the first and second ramp stage frames. The extension of the actuator unfolds the first stage of the ramp.

▶ The deployment of the ramp can be broken down into three steps. First, an actuator unfolds and extends the first stage of the bifold ramp. The release of the shoulder bolt and force of the cams initiates the full expansion of the ramp and platform from the sides of the robot. The result is a long, low inclined ramp with a long platform that can support two robots.

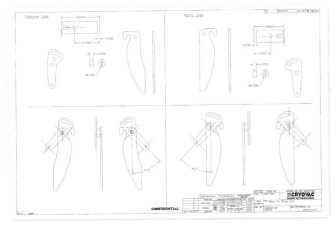

◀ When the actuator releases the shoulder bolt, a cam on each side is pushed to initiate the full deployment of both the ramp and platform.

▶ Design of the ramp/platform hold-release mechanism incorporates two functions. As the shoulder bolt is released, cams on each end of the actuator kick out the ramp and platform.

◀ Vertical legs support the extended ramp and platform to maintain level elevation under alliance robot weight. Rods connect to the robot base to push each leg into position as the ramp and platform are deployed.

▶ The robot designed by Golden Strip Flash easily accommodates two robots to score maximum bonus points at the end of a match. The folding design of the ramp and platform is controlled with actuators.

Elegant Elevation

Team 1448, the Parsons Vikings, implemented a design process that allows for ample creativity. Instead of designing a multifeature robot, they focused their energy and skills on one feature, making it as reliable and efficient as possible. Through approaching each match with a clear plan of action, this team executed an accurate and reliable performance every time.

METHODICAL DESIGN

Team 1448 employed a ten-step design process to methodically progress through the season. Each necessary stage, from team organization through construction and documentation, was identified to coordinate operations. The season began with the formation and organization of a design team. The game was reviewed, with the focus on scoring tactics. Preliminary sketches were critiqued, and the most feasible, valuable, and efficient design was selected for further development. Robot features were then drawn with AutoCAD to better visualize the layout, and prototypes were constructed to test for component functionality. Prototyping and refined drawings led to the manufacture of the final machine, and the team catalogued each mechanism through comprehensive documentation and graphic illustration.

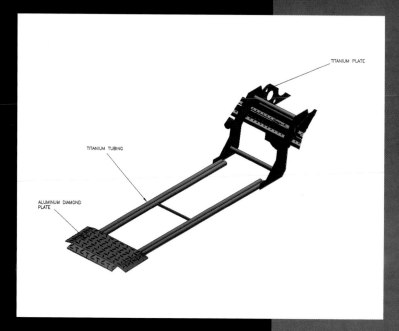

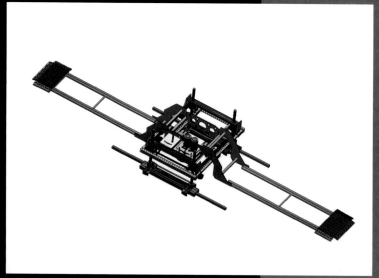

ONE MECHANISM, ONE FOCUS

Team 1448 was influenced by, and models their design process after, real-world industrial guidelines. The rules and restraints imposed by FIRST regulations are compared to size and cost specifications supplied by customers to designers. Game analysis revealed the different scoring potentials of game piece manipulation, autonomous activity, and alliance elevation. Local resources and manufacturing capabilities were assessed, and a design was chosen that could be supported by the available amenities.

Propositions of a four-wheel drive robot, an arm for game piece manipulation, and ramps for elevating alliance partners were abandoned. Strategy was simplified to include a single prominent function that would be proficient and reliable at a single task, rather than a machine that could perform many tasks with only a moderate level of success. The robot would incorporate two lifting arms, one on each side of the robot. These arms would reliably lift an alliance partner on each side, to secure bonus points at the end of each match.

Once the design of two lifting arms was chosen, the Parsons Vikings had to select a method of mobilizing them. The lifting motion could be supplied by pneumatics, but this was ruled out when research revealed that the maximum volume tanks allowed by FIRST regulations would not provide sufficient compressed air to lift the combined weights of two alliance partners and Team 1448's robot. A cable-and-pulley system was also considered to lift two platforms, but such a design required height and weight that the dimensional restrictions would not sanction. A rack and pinion design also could not comply with height and weight restrictions. The team settled on an arrangement that was both simple and small—a lead screw design powered by two CIM motors provided in the Kit of Parts.

TITANIUM ARMS AND ALUMINUM FEET

Each arm consisted of two parallel pieces of titanium tubing, with an aluminum diamond plate at the end to provide a slight incline to assist robots mounting the arms. These arms were attached to the robot with a custom-made titanium plate. Titanium was selected for these components because it provided both the stability and light weight required by the design. The low profile of the horizontal arms enabled most robots to easily drive over them and did not require alliance partners to navigate and climb a steep, bulky ramp. The alliance machines needed only to drive over the arms until reaching the blue bumpers.

The arms were hinged to fit within the size requirements at the start of each match. When the robot was in this drive mode, the arms remained folded while the robot played defense. When it came time to initiate the lifting process, the arms simply dropped down to the playing field floor.

While in drive mode, the robot maneuvered across the field with drive wheels mounted in the center of each side, with castor wheels at the four corners of the base for support. The location of the drive wheels enabled the robot to spin on its own axis. This agility, coupled with speed, helped the machine skillfully navigate the field to play defense against the opposing alliance.

In drive mode, the castor wheels made contact with the floor, while a foot assembly remained tucked up in the frame. When it came time to switch from drive mode to lift mode, this extruded aluminum foot replaced the wheels as a support. The foot was connected to the robot by four lead screws, located near each corner of the robot. Tucked inside the foot were stabilizers, which deployed from each end to provide more contact with the floor and distribute weight to facilitate lifting if the system was imbalanced due to the mounting of only one alliance partner. These stabilizers were deployed when holding pins, attached to the extrusion, lost contact with the robot frame as lifting began. Surgical tubing provided elastic tension to force the stabilizers to extend out of the foot.

▲ When the robot is in drive mode, it defends the opposing alliance from scoring on the rack. The arms are folded up and the stabilizers are stored in the feet.

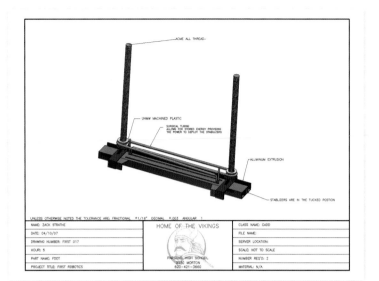

The foot assembly supports the robot when lift mode is initiated. The robot frame and wheels are elevated with the titanium arms as a single assembly.

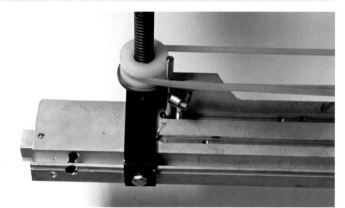

Stabilizers extend from the foot assembly to provide stability and increased contact with the floor. They enable weight to be unevenly distributed during lifting, so that a single robot can still be elevated.

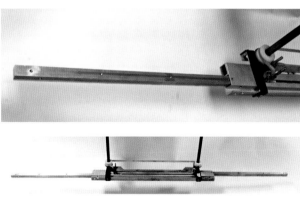

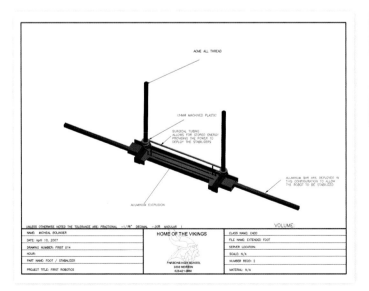

POWERING THE LIFT

A bracket attached the foot to the bottom of each lead screw. A series of bushings, bearings, spacers, and a sprocket was mounted to the extruded aluminum frame of the robot, between two fixed aluminum members. The lead screws fed through these components and the main robot frame. Two CIM motors were coupled with a custom gearbox and chain to provide power to the four lifting screws, while gear reduction enabled a faster lift time. When the robot was in lift mode, the sprocket was driven by the motors, which in turn rotated the lead screw. Because the sprocket was sandwiched between two fixed points, the turning of the screw caused the main robot frame to lift.

The arms were fixed to the main robot frame, so when the screws began to turn, the robot and arms elevated together. The four lead screws enabled the lifting of the Team 1448 robot and alliance partners. This design provided the robot with the ability to lift 450 lb. to a height of 17" (43.2 cm), well over the 12" (30.5 cm) height requirement for bonus points.

DEFENSIVE STRATEGY

This robot was programmed to perform during the fifteen-second autonomous period at the beginning of each match. When initiated, the robot rushed across the playing field, passing the rack to position itself between the opposing alliance and the spider legs. This defensive blocking move was effective when opposing robots were preprogrammed to score during the autonomous period. Every match played by Team 1448 at the Greater Kansas City Regional saw this machine charge across the field to reliably position itself in a defensive position. When the teleoperated period began, this robot was already on the opposing side, ready to confront and repel possible opponent scoring attempts.

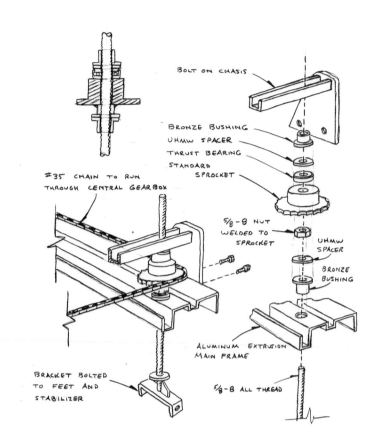

To lift alliance partners in the alliance home zone at the end of each match, ample space was required for deployment of the arms and a clear path for the robots to approach the arms. The robot was able to clear the area of stray game pieces by sweeping them out of the way with an extended arm.

With predictable yet effective autonomous programming and agile driving capabilities, Team 1448 created an imposing force for opposing alliances. The clean design of the titanium arms demonstrated an inventive approach to scoring bonus points, while the creative lifting mechanism enabled almost all robot styles to interact. Words cannot describe the fluent and seemingly effortless lifting of two 120-lb. machines. This robot captured both aesthetic and creative features in its elegant design.

▲ A concept sketch of the lead screw drive lift assembly shows how chain drives a threaded sprocket, attached to the robot frame, to raise and lower the robot.

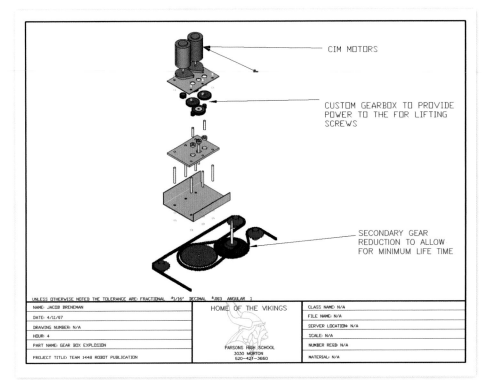

CIM MOTORS

CUSTOM GEARBOX TO PROVIDE POWER TO THE FOR LIFTING SCREWS

SECONDARY GEAR REDUCTION TO ALLOW FOR MINIMUM LIFE TIME

▶ Two CIM motors power the elevation of the robot. A gearbox and gear reduction provide the power and speed to lift up to 450 lb. to a height of 17" (43.2 cm) in fifteen seconds.

UNLESS OTHERWISE NOTED THE TOLERANCE ARE: FRACTIONAL ±1/16" DECIMAL ±.003 ANGULAR 1

NAME: JACOB BRENEMAN		CLASS NAME: N/A
DATE: 4/11/07	HOME OF THE VIKINGS	FILE NAME: N/A
DRAWING NUMBER: N/A		SERVER LOCATION: N/A
HOUR: 4		SCALE: N/A
PART NAME: GEAR BOX EXPLOSION	PARSONS HIGH SCHOOL 3030 MORTON 620-421-3660	NUMBER REQD: N/A
PROJECT TITLE: TEAM 1448 ROBOT PUBLICATION		MATERIAL: N/A

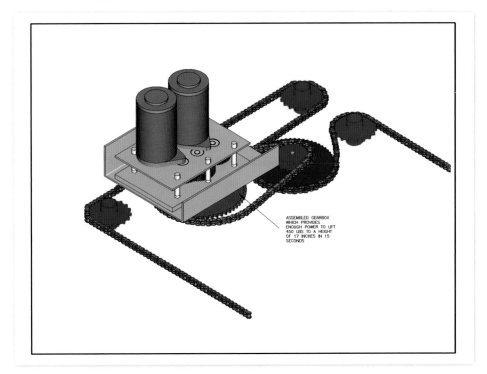

ASSEMBLED GEARBOX WHICH PROVIDES ENOUGH POWER TO LIFT 450 LBS TO A HEIGHT OF 17 INCHES IN 15 SECONDS

▲ The low-profile arms can elevate two full-sized robots. These arms enable alliance partners with low ground clearance or drive power to potentially score the bonus points.

▶ Once the arms are extended from Team 1448's robot, alliance partners need only drive over them and park to score bonus points. The robot vertically lifts itself and two alliance partners in one fluent motion.

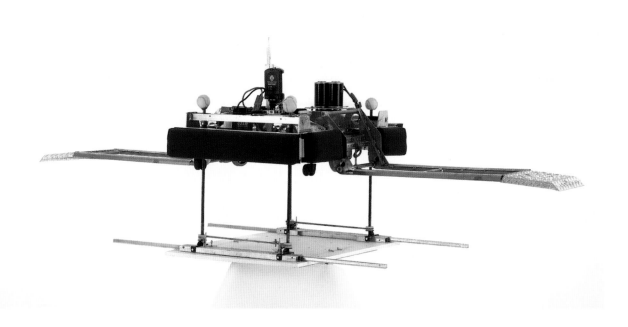

Elevation by Arc

Creativity is not necessarily the result of a spontaneous occurrence. It can be the result of brainstorming and careful planning, which provide the opportunity for the development of an uncommon solution. Team 1612, from Brooksville, Florida, exhibited their creativity in the design of a machine that promoted ease of assembly and troubleshooting. The unique ramp lift design was a particularly distinct feature. This robot was advantageous and functional, yet used minimal battery power and straightforward construction.

NARROWING DOWN RAMP DESIGN

The Robo-Sharks initiated the process of creating a competitive machine by establishing certain design criteria. The 2007 robot would feature ease of assembly and troubleshooting, functionality, and minimal use of power. From these guidelines, ideas were generated. The action on the playing field had the potential of being rough, so a design had to fulfill endurance testing and substantiation to constitute worthiness.

It had been established that the robot should include a ramp, so the team began to form possible design solutions. Some suggestions included a single-pulley system to raise the ramp, but this would not provide the necessary torque to raise heavy loads. Additional pulleys and extended pivot points could improve the power of the system, but they would not fit within robot size constraints.

▲ SolidWorks is used to design an aluminum component. Drawings such as these enable team members to clarify theoretical models and verify feasibility.

Brainstorming created many possibile solutions that were discussed by the team. The ideas presented were narrowed down to five possible ramp sketches, with consideration of how each design would function during play. These five proposals then underwent a full design review and were ranked. Criteria considered included ease of construction, minimal parts, and cost. From these five, the design was further narrowed down to two distinct machines.

The two selected ramp layouts were prototyped and tested to determine the best final design to transfer to the robot. The process of first narrowing the design options down to five yielded time saved on selecting solutions. This time budgeting also drove the narrowing of designs down to two, which allowed for material conservation because only two prototypes were constructed.

Drawings of the two prototypes progressed from rough hand sketches to AutoCAD and SolidWorks drawings. Three to four drawings were made for each of the two prototypes, to elucidate concepts for further discussion and elimination.

The selected design incorporated a pneumatically powered ramp that could be raised and lowered, elevating alliance robots for bonus points during the end game period. Two independent arc shapes were hinged underneath the ramp. An actuator was attached to the end of the arc not connected to the ramp. Extension and retraction of these actuators imparted the raising and lowering motion of the ramp.

▲ A prototype of arcs that are pulled to raise the ramp is dismissed after analysis of pneumatic operation and friction determines that a pushing method is more efficient.

▶ The ramp concept is tested using a wooden prototype. The slope of the ramp must be gradual enough for alliance robots to mount without excessive effort.

IMPLEMENTING ARCS FOR ELEVATION

The first prototype of this ramp and arc design introduced the question of whether it was more efficient to push or pull the arcs to raise the ramp. The only difference in construction was the orientation of the arcs, so both methods were tested. Each worked, but review of the pneumatic system schematics and friction analysis determined that it was more efficient to push the arcs to raise the ramp.

The arcs were made of high-density polyethylene (HDPE). Different sizes of arc were tested to find the ideal outer radius for leveling the ramp. An arc with an outer radius of 12" (30.5 cm) was first applied, but the raised ramp resulted in a platform that was too low. A 16" (40.6 cm) outer diameter arc was tested next, but the resulting platform was too high. Finally, a 13" (33 cm) outer radius arc positioned the ramp at the same height as the main body of the robot, for a smooth, level surface. A custom-made clevis attached each of the two pneumatic rods to an arc. They were specially designed to conform to the arc position when it was both stowed and deployed. The U-shaped piece bolted onto a pin running through the arc. The arcs were attached and hinged to the frame of the ramp with a triple nut and a single friction washer. These, along with an application of Loctite adhesive on the threads, prevented the arcs from working loose after repeated deployments.

Custom-made ramp-offset brackets were made to assist the folding of the ramp and prevent binding. A spring placed under tension created a counterforce and provided a smooth upward motion as the ramp was raised. Tread grip tape covered the surface of the ramp to provide traction.

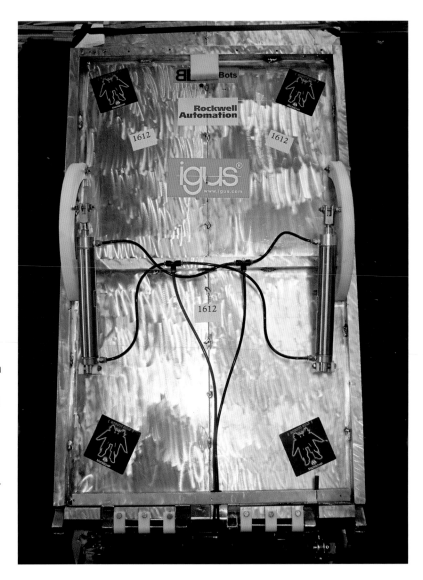

▲ An actuator and arc are located on each side underneath the ramp. A custom-made clevis connects each pneumatic rod to a pin running through the arc.

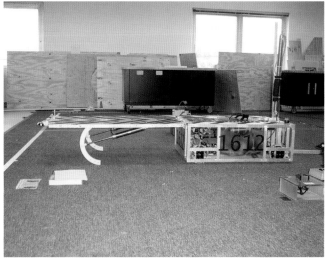

▲ Once the ramp is deployed, alliance robots can drive up to the platform. Black tread grip tape provides added traction. Activation of the two actuators under the ramp elevates it to an even level with the platform, above the 12" (30.5 cm) height requirement for bonus points.

▶ The ramp is securely connected to the robot frame with custom-made ramp-offset brackets. These brackets prevent binding during folding and a tension spring enables smooth motion. They are seen here from the top of the ramp and in a close-up view under the ramp.

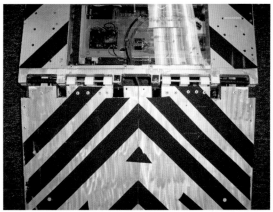

▲ A student sets up the CNC router to fabricate the arcs that will lift the ramp. Auto-CAD drawings are transferred to a program that created tool paths to correctly cut the building material.

◀ The two arcs support the raised ramp. They are hinged at one end to the ramp frame and are connected to the actuators in the center. Extension of the actuators pushes, rather than pulls, the arcs into a raised position.

▶ The incline angle allows even robots with large bumpers to climb the ramp. The platform could support one robot, and the ramp could lift and support a second robot, with ample room. This ability can score up to sixty bonus points per match.

PNEUMATIC DESIGN AND SIMPLIFICATION

Possible arrangements for the pneumatic system were investigated to produce the desired ramp movement. The pneumatics manual provided by FIRST included a chart relating pressure, bore size, and output forces of both extension and retraction motions. Team 1612 used this information to verify that the conceptual system would fulfill strength and power requirements. The minimum and maximum pressure ranges were calculated, along with the distance the assembly needed to travel. From the acquired data, a wooden prototype was constructed on which the pneumatics could be mounted and tested. AutoCAD drawings of the components were saved in an interoperable format and transferred to spectraCAM Milling software from intelitek. This software generated tool paths from the part drawings, capacitating the milling of parts with a computer numerical control (CNC) milling machine and CNC router.

The air for the pneumatic system was initially provided by a compressor, which had been included in the system testing and would be mounted on the robot. The compressor added considerable weight to the robot, and was not necessary, as the air supply in the storage tanks did not need to be replenished during a match. The *2007 FIRST Robotics Competition Manual* stated that a robot could start each match with prestored energy. The design was modified to remove the compressor from the robot and enable easy access to the system for prematch charging of a maximum of 60 lb./sq. in.

The resulting robot included a pneumatically powered ramp that could lift 180 lb. to a height of 13'' (33 cm) above the field floor and a platform that could support a second robot. Team 1612 used a minimal amount of tubing, wiring, and fasteners to simplify assembly and disassembly for troubleshooting. The Robo-Sharks credit their creative ability to the narrow build deadline. It is when under pressure that original ideas, such as the ramp-lifting arcs, emerge. The focus of energy on rapid and sensible design solutions undeniably impelled the team's accomplishments.

▶ With its ramp folded up, the robot is ready to begin a match. Cold cathodes are mounted in the undercarriage to glow either red or blue, depending on the assigned alliance color for each match.

FIRST: Objects, Individuals, and Societies
Woodie Flowers

FIRST is a many-faceted thing. This book is primarily about the objects of FIRST.

Behind each of the robots, one can find at least two layers. One layer is related to the individuals, the other to the social interaction around the robot's creation. All three—the objects, the person, and the societies—are very important.

I believe the powerful interaction of these layers is one of the strengths of FIRST. The robots result from an intimate dance with Mother Nature. Absolutely objective, she applies all her rules all the time. Under no circumstances can one receive even a momentary reprieve. If the movie running in the mind of the robot designer is not consistent with Mother Nature's notions, the robot will not do what the designer imagined. That objectivity is applied without sympathy or emotion.

Through that cool, objective interaction with objects, the designer/inventor/student receives unmediated feedback. If it works, it works, and the intimate dance becomes lovely and satisfying. If a design does not work, one can give up or learn more about creative redesign and problem solving. The robot designers in this book did not give up but learned through repeating the cycle until the dance worked. They benefited from a kind of rational evaluation important to developing self-esteem.

If they were lucky, the designers also received ample feedback about their personal contributions and their interaction with their teams and with society. The FIRST soup includes "it," "me," and "us." The best recipes for that soup take full advantage of all three ingredients. If one is challenged by a stretch goal, such as building a sophisticated machine with a group of colleagues while at the same time behaving in a gracious and professional manner, that is a great opportunity to develop the kind of robust self-image essential to being a contributor to society. If we believe we "can" based on a mix of objective evaluations by nature and subjective feedback by other individuals, and by society as a whole, we probably "can."

Earned self-esteem is robust and strong enough to serve a lifetime.

▶ Dr. Woodie Flowers is the Pappalardo Professor Emeritus of Mechanical Engineering at the Massachusetts Institute of Technology, a Distinguished Partner at Olin College, and cofounder of the FIRST's cornerstone program, the FIRST Robotics Competition. Professor Flowers is chairman of the FIRST Executive Advisory Board and participates in the design of the FIRST Robotics Competition game each year. He has served as a national advisor to the FIRST Robotics Competition since its inception.

DIRECTORY OF AWARD WINNERS

DELPHI DRIVING TOMORROW'S TECHNOLOGY AWARD

TEAM 118: NASA-JSC and Clear Creek ISD
League City, TX

Eric Ambrose, Andres Caliz, Jose Cantu, Jeff Ciabattoni, Sunayana Chopra, Steven Cooper, Will Cover, Patrick Creamer, Christine Cucinotta, Pavan Dave, Matheus Fernandes, Ithamar Glumac, Robbie Greene, Eric Griffis, Nate Harper, Jacob Johnson, Tom Johnson, Jason Klein, Josh Kuhn, Remi Kutowy, Stanley Lewis, Trent Lilly, Scott Longmore, Solomon Mathew, Matt McCann, Michael McGee, Micheal Merta, Zoheb Noorani, Nick Pajot, Scott Peck, Sarah Propst, Zachary Russell, Christine Salvo, Matthew Schoonover, Abigail Sprague, Kaylynn Burns, Teresa St. John, Lora O'Neil, Rob Ambrose, Lucien Junkin, Jonathon Lutz, Lyndon Bridgwater, Nathan Fraser-Chanpong, Kris Verdeyen, Chris Sanborn, Nathan Smith, Roger Rovekamp, Justin Ridley, Tom Ebersole, Don Nolan, Mike Valvo, Reyes Vega, David Glover, Bill Bluethmann

TEAM 173: CNC Software, JP Fabrications, Nerac, Inc., United Technologies Research Center, East Hartford High School, Rockville High School, and Tolland High School
East Hartford, CT

Carl Berner, Travis Bush, Aleena Cusson, Geoff Eckert, Steve Fielding, Courtney Hanson, Adam Holt, Justin Holt, Josh Houle, Daisy Karas, Chris MacCoy, Emily McDermott, James McDermott, Brittany Mullen, Lindiwe Ngobeni, Rachael Nystrom, Mike Pelisari, Josh Perry, Stephen Rizzo, Steve Sanzo, Jamie Suchecki, Evan Thorpe, Ben White, Trevor Krechko, Jon Kentfield, John Rizzo Jr., Jeff Rodriguez, R. J. Sisca, Bill and Bridgette Berner, Laura and Tom Bush, Paul and Maire Cusson, Connie Ekstrom, Bruce Hockaday, Jen Holt, Roxanne Houle, Dave MacCoy, Margaret McDermott, Andy Moiseff, Charles Nystrom, Jim and Pat Palmer, Jeff and Lisa Perry, Kim and John Rizzo Sr., Sal Saitta, Sal and Donna Sanzo, Alice Sisca, John Suchecki, Robert Triggs, Michael White

TEAM 375: Richmond County Savings Foundation, Verizon, Con Edison, and Staten Island Technical High School
Staten Island, NY

Dan Hoizner, Eric Demaso, Michael Iannelli, Samantha Catanzaro, Michael Giehl, Steve Scandaglia, Rich Boyle, Daniil Zolotarev, Eugene Shteynvarts, Steve Prokopenko, Graham Freedland, William Tsimerman, David Doh, Vivek Laungani, Michael Siegel, Dennis Giurici, Steve Raile, Alan Bailey

TEAM 1086: Henrico Education Foundation, Fexicell, Qimonda NA, McKesson Corp., Midwest Industrial Chemical Co., Showbest Fixture Corp., and Deep Run High School
Glen Allen, VA

Connor Clarke, Fran Nolen, Hans DeKonig, Thomas Halish, Dave Cohen, Dave Underwood, Michael Young, Vikram Rajasekaran, Colin Windmuller, Brian Rutherford, Kevin Pfab, Alex Curfman, Christina Walinski, John Kutz, Nick Arancibia, Bobby Anderson, Nick Zelinske, Thomas Owen, Matt Cotton, Anthony D'Alessandro

TEAM 1985: NASA, Emerson, DRS, UGS, and Hazelwood Central High School
Florissant, MO

Cathy Sylve, Tom Wendel, Mike Duello, Dan Robinson, Mike Quigley, Lauren Hart, Jeff Cosper, Lindsay Wendel, Tyler Saffer, Jarrod Duello, Justin Duello, Zachary Robinson, Tyler Robinson, J. D. Jones, Dan Forbes, Danielle Sylve, Chloe Cotton, Garrick Johnson II, Mike Cooper, Scott Pine, Scott Cosper, Shawn Bodden, Jake Gardner, Ryan McHugh, Brett Wallace, Brittany Norwood, Taylor Hyndman, Ota Kukral, Garrett Hemann, Sean Hibbits, Michael Braun, Jesse Gill, Ranell Cavitt, Robert Harrell, Blake Hiza, Tommy O'Brien, James Wilkinson

TEAM 1987: Ewing Marion Kauffman Foundation, Universal Plumbing, R&D Tool and Engineering, Kastle Grinding, High Tech Laser & Engraving, Weber Carpet, Burger & Brown Engineering, CBW Automation, Spring Dynamics, HS Springs, Permatex, W. M. Berg, Leslie's Pool, Wolverine Fluid Power, Inc., Kastle Manufacturing LLC, Price Chopper of Lee's Summit, Carbon Fiber Tube Shop, Lubriplate, Honeywell, RSC Equipment Rental, and Lee's Summit North High School
Lee's Summit, MO

Steven Armour, Sarah Bringer, Jacob Dobyns, Dylan Edwards, Craig Eubank, James Freeman, Becky Hand, Emma Hand, Nathan Raymond, Reed Ross, Haley Russell, Alan Salkanovic, Laura Shafer, Zac Sloan, Logan Smith, Shane Smyth, Max Steiner, Lee Swaim, Luke Turner, Tom Reynolds, Jackie Campbell, Tom Hand, Rex Luchtel, Murray Ford, Randy Dobyns, Steve Penrod, Gary Shafer, Dan Phillips, Mark Brown, Jeff Grisham, Chuck Stephenson, Candice Stephenson

GENERAL MOTORS INDUSTRIAL DESIGN AWARD

TEAM 25: Bristol-Myers Squibb and North Brunswick Twp. High School
North Brunswick, NJ

Shawn Ahmad, Bria Alexander, Kiran Aravapalli, Srirag Babu, Michael Bauer, Matthew Beckles, Kristian Calhoun, Tina Ciance, Michael Ciance, Antoine Dorsey, Richard Ebright, Lauren Fackelman, Morgan Gillespie, Kirsten Guevarra, Nitesh Harid, Jie Huang, Raj Kannappan, Jatil Kodati, Bryant Lam, Christopher McLean, Shaun McNulty, Brian Mollica, Mohit Moondra, Nishsit Nandankar, Shane Ogunnaike, Aditya Pandyaram, Nisha Parikh, Neil Parikh, Dev Patel, Hannah Petersen, Nicholas Ramos, Amy Rankin, Mike Savin, James Schroeder, Jigar Shah, Lawrence Shen, Robert Slattery, Stephanie Solis, Alexa Stott, Daniel Tang, Dhananjay Thanikella, Rishi Tirumala, Eric Veguilla, Maria Xu, Charley Smaltino, Walter Suchowiecki, John Dusko, Tony Kukulski, Mike Lubniewski, Kevin Durham, Paul Kloberg, Wayne Cokeley, Bob Goldman, Mike Palazzo, Roger Weiss, Lauralynne Cokeley, Bruce Ciance, Beverly Stott, Mike Savin, Lynne McLean, Michael Ciance, Shaun McNulty, Bharat Nain, Tom McLean

TEAM 75: J&J Consumer and Personal Products Worldwide and Hillsborough High School
Hillsborough, NJ

Samantha Abramovitz, Amanda Albin, Ian Berezansky, Laura Berger, Katie Bianchini, Victoria Biegel, Andrew Blackburn, Nick Bottenus, David Boyea, Adam Brody, Alex Burman, Brittany Capalbo, Jenny Cattron, Eric Chang, Jimmy Chapuran, Randy Coddington, John Czernikowski, Lisa Czernikowski, Alison DeMaio, Manny DeMaio, Thomas DiPalma, Bryant Domitrowski, Andrew Douglas, Rebecca Doyle, Alison Duch, Brandon Eberhardt, Nate Fong, Jim Gibbons, Katherine Glass-Hardenbergh, Tim Glowinsky, Arek Gnap, Ricky Grandell, Stephanie Greenspan, Amanda Grek, Joanna Hanson, Alex Kimmelman, Sushanth Kodali, Matthew Lancellotti, Bo Li, Yihua Lou, Alex MacKinnon, Tim McCarty, David McDonald, Sarah Mitrani, Scott Newman, Carol Petrosyan, Dan Rapacki, Chris Salvito, Dave Schramm, Matthew Simpson, John Skeele, Jon Sparrow-Hood, Ryan Stainken, Justin Stockman, Matthew Stockman, Rohith Surampudi, Adam Thackara, Bryan Toth, Nick Urban, Jon Wisniewski, Seth Zubatkin, Michele Zubatkin, Matthew Chernoff, Robert Bianchini, Theresa Edwards, Fred Hartman, Mari Hou, Rich Prospero, Carmine Rizzo, Don Wells, Tamer Soliman, Paul Lydick, Ron Biegel, Jason Sacco, John Fabian, Tammy Park, Giselle Sulit, Ron Grasso, Eric Lunde, Michael Luedtke, Steve Tricarico, Tara Zedayko, John Douglas, Michele Zubatkin, Steven Zubatkin, Irene Czernikowski, John Czernikowski, Dave Blackburn, Ben Fong, Annette Fong, Alan Kimmelman, Alex Toth, Debbie Boyea, Virginia Skeele, Walter Sparrowhood, Mary DiPalma, Marie Stainken

TEAM 111: Motorola, Rolling Meadows High School, and Wheeling High School
Schaumburg, IL

Raul Olivera, Dan Posacki, Dan Rooney, Suren Devarashetty, Kirstie Cannon, Conor Delaney, Lisa Dohn, Steven Heer, Tyler Idzerda, Rebecca Leung, Maxwell Stehmeier, Brian Werling, Dave Scheck, Mike Soukup, Dave Flowerday, Kristin Schultz, Erik Edhlund, Nathan Harman, Brian Kuczynski, Greg Laprest, Jacob Mandozzi, Michael McMinn, Christopher Moss, Staci Tipsword, Vladimir Voskoboynikov, Adam Wojcik, Nate Troup, Al Skierkiewicz, Michael Anderson, Ryan Klaproth, Josh Koci, Stelio Kraniotis, John Kuehne, Stephen Mikels, Kent O'Neill, James Shawley, Karl Swanson, Katherine Thompson, Miten Champaneri, Julie Albrecht, Mark Rooney, George Graham, Dan Green, Bruce Johnson, Ryan Kerekes, Phil Rybarczyk, Sharon Rybarczyk, Tom Thompson, Mark Faust, Eric Faust, Dottie Skierkiewicz, Pam Balcer, Mark Koch, Tim Patterson, Amy Vannatta, Brian Allred, Luke Dahlgren, Joseph Doose, Jeremy Evers, Mark Faust, Aaron Glover, Susanne Groen, Kelly Hayden, Thomas Koehler, Brian Korves, Daniel McMahon, Jason Park, Parth Patel, Teagan Russell, Alex Sabatka, Paul Simon, Andy Tatkowski, Anthony Thompson

TEAM 148: RackSolutions.com, L-3 Communications Integrated Systems, and Greenville High School
Greenville, TX

Dustin Beasley, Bethany Follett, Nathan Follett, Ben Gilbert, Lane Gould, Ethan Isham, Trent Jones, Keith Kaden, Austin Lambert, Julie Latimer, Halley Lawson, Mark Mahrer, Logan Marcum, Zack McClellen, Kory Porter, Adam Risley, Justin Tharp, Will Thornton, Clayton Torrance, Stacey Walker, Melanie Wisdom, Karlee Wright, Chris Carnevale, Chris Follett, John Hodapp, Ben Kolski, Brandon Martus, Steve Maxwell, Vanessa Pope, Adam Reppond, Kevin Rodgers, Ken Stroud, Johnny Tharp, Ricky Torrance, John V-Neun, Casey Welch, David Coleman

TEAM 330: J&F Machine, NASA-JPL, Raytheon, and Hope Chapel Academy
Hermosa Beach, CA

Sam Couch, Eric Husmann, Elijah Mounts, Richard Norris, Shane Palmerino, Dakota Payton, Courtney Roberts, Matt Rose, Caitlin Steele, Faith Steele, Larry Couch, David Drennon, Matt Driggs, Steve Eccles, Tom Freitag, Chris Husmann, Michael Palmerino, Ben Roberts, Ric Roberts, Andy Ross, Gregory Ross, Joe Ross, Penny Ross, Robert Steele, Zach Steele, Rick Varnum

TEAM 1126: Xerox Corporation and Webster High School
Webster, NY

Ralph Hudack, John Gramlich, Bob Schlegel, Ray Bassett, Mike Bortfeldt, Karen Buck, Ken Buck, Carol Milton, Mitch Milton, David Arena, Bob Mintz, Randy Napoli, Ryan Melnychuk, Scott Lockhart, Bryan Bayer, Wendy Marsh, Leon Cormier, David Schenk, Courtney Miraglia, Dante DiLella, Dan Capizzi, Cory Gramlich, Daniel Arena, Chris Armbruster, Justin Bassett, Alex Bolognesi, Bob Fekete, Timothy Francis, Laura Garland, Dylan Gramlich, Tyler Holroyd, David Johnson, Josh Metcalf, Nic Milton, Zachary Mintz, Sam Rueby, Mitchell Sedore, Jeremy Tice, Danielle White, Chris Bayer, Chris Brown, Ryan Chang, Matt Dentico, Meghan Dwyer, Danielle Fox, Sean Goodrich, Matthew Heid, Jessica Kane, Kenny Miller, Nick Miraglia, Tori Parry, Alejandro Rivera, Ian Teague, Nick Volo, Mike Dwyer, Bev Gramlich, Sue Mintz, Karen Holroyd, Irene Bayer, Mike Miraglia, Dale Boudreau, Michelle Bolognesi

DIRECTORY OF AWARD WINNERS

MOTOROLA QUALITY AWARD

TEAM 100: PDI Dreamworks, SRI International, A Better Mousetrap Machining, Woodside High School, and Carlmont High School
Woodside, CA

Janet Creech, Arlene Kolber, Laura Robeck, Peter Adams, Andrew Baltay, Rachel Cape, Nick Chang, Billy Chartain, Camryn Douglas, Geoff Dubrow, Luke Gray, Johnathan Greer, Shawn Housholder, Kevin Irish, Troy Lemon, Henry Lewis, Kenneth Phung, Michael Rhodes, Steven Rhodes, Hannah Sarver, Brad Saund, Arin Singh, Trevor Sturm, Martin Taylor, Andrew Theiss, Haley Valletta, Alex Waschura, Chris Wuydts, Danny Aden, Richard Burgess, Jim Cape, Susan Dubrow, Laura Rhodes, Lindsay Steinfeld, Bob Steinfeld, Kristine Taylor, Al Theiss, Janice Valletta

TEAM 190: Worcester Polytechnic Institute and Massachusetts Academy of Math and Science
Worcester, MA

Elizabeth Alexander, Michael Carkin, Danielle Jacobson, Dan Jones, Steve Kaneb, James Kirk, Andy Lewis, Annie Lin, Wayne Liu, Josh Matthews, Andrew Nehring, Matt Netsch, John Nsubuga, Dan Praetorious, Colin Roddy, Raisa Trubko, Brad Miller, Ken Stafford, Diana Berlo, Kevin Bobrowski, Mike DiBlasi, Aaron Holroyd, Brian Loveland, Erik Macchi, Evan Morrison, Ciaran Murphy, Francis O'Rourke, Ruth Toomey, Sabrina Varanelli, Paul Ventimiglia, Alex Volfson, Christopher Werner

TEAM 234: Allison Transmission, Rolls-Royce, 74 Proud Grandmas, and Perry Meridian High School
Indianapolis, IN

Jon Anderson, Auriel Aurellano, Daniel Biehl, Brice Brenneman, Warren Brewster, Jamie Cartwright, Anthony Chastain, Brian Clegg, Ninah Clegg, Megan Colwell, Alec Curtis, Matt Deal, Rachel Dobbs,

Sam Gadbury, Miranda Goelz, Matt Marchione, Ben Martin, Matt Martin, Mitch Mounts, Nahan Nitsch, Sohaib Omari, Alan Raynes, James Roach, Corey Rohl, Rachel Schrage, Amrit Singh, Jared Smith, Emily Stephens, Kevin Stumpf, Breanna Tonte, Joe Tonte, Brandon Walsh, Phillip Welsh, Mark Dobbs, Chris Fultz, Ed Henry, David Kelly, Kevin Kelly, Mike Koiro, Dane Rodgers, Larry Stumpf, Steve Warner, Scott Ritchie, Steve Wherry, Norm Chastain, Kelli Fultz

TEAM 384: GE Volunteers, Qimonda of Richmond, Flexicell, ShowBest Fixture Corp., Specialty's Our Name, ChemTreat, Sam's Club, CAPER, ITT Technical Institute, Henrico Co. Education Foundation, and Tucker High School
Richmond, VA

Aravind Kandalam, Carol Cotton, Chris Banfield, Driver Branson, Gigi Chang, Jordan Lowell, Grayson Atkinson, Greg Cotton, James McCarson, Jeff Atkinson, Jim Schubert, Joseph Matt, Josh Griff, Juanita Branson, Julie Norris, Leo Meire, Mae Cinco, Marshall Turner, Mary Hutchinson, Nadean Cubero, Patrick Cole, Robert Benway, Terry Daniels, Will Gathright, Connor Clarke, Dave Underwood, Thomas Halish, Dave Cohen, Jiten Narang, Chris Hockaday, Danny Nichols, Galen Nickey, Kien Tran, Jay Gowda, Jackie Murrell, Mitchell Cisney, Meghan Cook, Mary Jackson, Rameez Khimani, Robert Edgeworth, Rick Boogher, Rick Lewis, Scott King, Steven Hutchinson, Vivian Chan, Annie Denison, Brad Wilson, Amber Manche, Sarah Keener, Tara Lahens

TEAM 1114: General Motors—St. Catharines Powertrain, Allied Marine and Industrial, and Governor Simcoe Secondary School
St. Catharines, ON

Matthew Coffey, Lindsay Davies, Melissa Doornekamp, Sam Engstrom, Candice Evans, Ian Froud, Jesse Graham, Chris Lyddiatt, Stuart MacGregor, Zak Mason, Keith Millar, Brandon Pruniak, Jessa Pruniak,

Ryan Shaw, Nik Unger, Luke Visser, Denise Wong, Geoff Allan, Jeff Beckett, Derek Bessette, Emerald Church, Michael DiRamio, John Froud, Steve Hauck, Tyler Holtzman, Karthik Kanagasabapthy, Esther King, Eric LePalud, Ian Mackenzie, Don Mason, Catherine McNamara, Joanne Pruniak, Mike Pruniak, Ben Radtke, Steve Rourke, Brent Selvig, Garry Unger, Greg Phillips, Todd Willick

TEAM 1714: NASA, Rockwell Automation, Quad Tech, Pentair Water, Marquette University, Society of Plastic Engineers, MSOE, American Acrylics USA LLC, and Thomas More High School
Milwaukee, WI

Michael Wittman, Bob Grilli, Jon Anderson, Mark Fons, Kevin Kolodziej, Jeff Krueger, Art Laabs, Dan Lang, Ron Peter, Jim Rehfeldt, Mark Rivest, Tom Silman, Chris Stack, Bill Stoltenberg, Louis Verstegen, Mary Jo Wittman, Sarah Wittman, Matt Kallerud, Karl Akert, Beth Burton, Patrick Carroll, Jamie Coyne, Megan Czaplewski, Joseph Grilli, Dustin Holzhauer, Jeff Krawczyk, Joey Laabs, Ken Leung, Steven Lynch, Mara Mamerow, Nathan Peter, A. J. Rehfeldt, Jenni Rehfeldt, Abby Schaefer, Nate Schaefer, Matt Schultz, Mary Sowinski, Jared Verba

ROCKWELL AUTOMATION INNOVATION IN CONTROL AWARD

TEAM 33: DaimlerChrysler and Notre Dame Preparatory School
Auburn Hills, MI

Alex Scales, Amanda Suiniak, Andrew Poisson, Andrew Woodcox, Brian Fain, Carla Spicuzzi, Caroline Beyer, Chet Fleming, Chrissy Kaub, Connie Pulliam, David Huston, David Wulbrecht, Devon Stuart, Isaac Rife, Jason Monroe, Jerry Yahmatter, Jim Zondag, Joey Navare, Jon Shepard, Keith McDonald, Kevin Thompson, Mike Hergert, Nick Wrobel, Ron Culbert, Ryan McIntosh, Sarah Huston, Sarah Yahrmatter, Sean Grogan, Tim Grogan

TEAM 40: intelitek/BAE Systems / I.P.Y.M. and Trinity High School
Manchester, NH

Dan Larochelle, Joe Pouliot, Jessica Boucher, Karen Pringle, Adam Martin, Dennis Tappin, Mike Gleason, Elena Ainley, Dan Brady, Dan Breault, Kiara Feliz, Matt Coryea, Sean Dixon, Alex Gadecki, Jon Gaffen, Matt Gleason, Adam Golding, Colleen Hug, Jaz McDowell, Tim Moreau, Brendan Newcott, Emily Ogilvy, Hanna Ogilvy, Jeff O'Rourke, Tyler Pepin, Robert Pitt, Matt Rotier, Emily Siemiesz, Megan Uberti, Carlo Victorio, Matt Walsh, Erica Wong, Jessica Wong

TEAM 386: Harris Corporation, Florida Institute of Technology, BD Systems, Compass Solutions, Ace Hardware, EDAK, Ascent Media, Brevard Public Schools, Melbourne High School, Satellite High School, West Shores Junior High School West, Shores Senior High School, New Covenant Christian School, and Home Schooled Students
Melbourne, FL

Allie Armstrong, Alex Bailey, Kimberly Beck, Timothy Bell, Dana Connor, Brandon Couts, Kimberly Day, Rose Deffenbaugh, Jesus Diaz-Rivera, Joshua Eberle, Alex Erhan, Emily Gilger, Josh Goldfarb, Charlie Haas, Megan Hubbard, Danny Kemp, Riley King, Adam Lindsley, Nate Longbons, Eric Lyons, Kyle Lyons, Aanisah McClendon, James Mussman, Dane Newton, Samantha O'Brian, Fred Pitten, Daniel Risner, Nathyn Smith, Nathan Spoly, Chirstopher Struttmann, Katie Touchton, Aaron Watson, Zachary Winton, Trent Wood, Enrico Pischiera, Tom Gabeler, Juan Davila, Lynn Deffenbaugh, Okie Baughman, Joe Bernier, Ron Couts, Mike Dion; Jim Fisher, DeAnn Sperber, Don Davis, John Day, Richard Werner, Jim Lyons, Diana Touchton, John Bailey

TEAM 418: National Instruments, BAE Systems, Inc., LASA Robotics Association, and Liberal Arts and Science Academy of Austin
Austin, TX

Anthony Bertucci, Alexander Yau, Lewis Henrichs, Elaine Lee, Danny Diaz, Ethan Reesor, Luis Cantu, Jenny Lau, Jane Young, Gabe Ochoa, Mitchell Wilkinson, Scarrlett Kimmett, Randy Baden, Henry Mareck, Natalie Bixler, Sam Wolfson, Jim Mareck, John Mareck, Omneya Nassar, Jay Kapoor, Coy Fancher, Jennifer Gibbons, Riad Nassar, Arash Saeedi, Allen Smith, Gerry Salinas, Price Vetter, Christina Brazell, Clint Olmos, Judy DeWitt, Ryan Newton, Anthony Lamme, Cruz Monrreal, Nick Steinhauser, Alan Xu, David Riffey, Krista Nicklaus, Lokesh Anand

TEAM 1629: Beitzel Corporation and Garrett County Public Schools High School
McHenry, MD

Philip Adams, Andrew Beitzel, Dylan Beitzel, Brian Bramande, Devynn Brant, Sarah Coddington, Mitch Hall, J. P. Law, Stephanie Lee, John McGettingan, James Mullenax, Emily Rosser, A. J. Storck, Sarah Storck, Zach Trautwein, Chris Wood, Titus Beitzel, Donnie Beitzel, Jed Crawford, Larry Friend, Cheryl Gnegy, Josh Hinebaugh, Kearstin Hinebaugh, Lisa Malone, Phil Malone, Larry Mullenax, Frank Shap, Chuck Trautwein, Erik Wood

TEAM 1731: NASA, Raytheon, Freestate Electronics, King, Lam and Fresta Valley Christian High School
Marshall, VA

Leah Bode, Josh Buckley, Sarah Buckley, Kyle Fox, Chrissy Hohenberger, Daniel King, Levi Leppke, Lindsay Leppke, James Minihan, David Ross, Andrew Stalker, Peter St. Jean, Matthew Yeager, Woody Bode, Daniel Buckley, Jim Fox, Ron King, David Lam, Donovan Lam, Brent Leppke, DeWayne Leppke, Jim Minihan, Robert Stalker

DIRECTORY OF AWARD WINNERS

XEROX CREATIVITY AWARD

TEAM 79: Honeywell Inc. and East Lake High School
Clearwater, FL

Stephen Chadwick, Ryan Cianciolo, Katt Gera, Candace Hazelwood, Ryan Heaphy, Adam Horne, Samantha Hunter, Even Kardiff, Taylor Kuizenga, Ken Lindermana, Chris Martin, Aaron O'Hare, Evan Prado, Nathan Roncal, Andrew Sansing, Elisha Stevenson, Mark Wahnish, Craig Zeitlin, Jay Zuerndorfer, Kayla Williams, Jakob Christopher, Matt Azarian, Kelly Powell, Ben Sansing, Michael Strong, Brian Yi, Sierra Reshaw, Danny LaMont, Dan Hamillon, David Thibodeau, Zach Anderson, Carolyn Voytlan, Michael Habashy, Josh Harrelson, Mike Bilello, Jonathan Vlastares, Drek Roncal, Kellie Brooke, Bryan Maynard, Ben Peck, Jeff Cron, Danny Gorman, Stephanie Sims, Alex Sims, Brian Camp, Ron Hartman, Kim Heinicka, Keith Hildebrand, Barry Leedy, Shelley Arnold, Marco Lombardo, Jerry Waechter, Tom Filipek, Ken Gardner, Dave Hollingsworth, George Donaldson, Linda Wahnish, Paul Wahnish

TEAM 357: PECO Exelon and Upper Darby High School
Drexel Hill, PA

Joe Troy, Beth Hale, Adam Schuman, Mike Crane, Christy Troy, Megan Durkin, Jules Scogna, Chris Gerlach, Joe Adams, Matt Blubaugh, Sid Joshi, Pat Kneass, Jenn Knowlton, Lauren Campion, Melissa Cell, Jeff Davis, Ben McCarron, Garrett Sapsis, Chris Manuel, Dan Troy, Andrew Adaman, Kevin Durkin, Christina Ghilardi, Joe Gro, Jon Knippschild, Bob Knowlton, Rachel McKlindon, Tony Scala, Glenn Baker, Kelly Leonard

TEAM 1100: Rohm and Haas and Algonquin Regional High School
Northborough, MA

Dan Auger, Dan Baker, Paul Baker, Ray Belanger, Andrew Bigelow, Tom Castelli, Alex Cheung, Dan Church, George Clarke, Faith Clayton, Benjamin Croteau, Steven Crowe, Tom Estelle, Jimmy Foley, William Frankian, Robert Galgano, Ben Gibbons, Hardy Hartwell, Daniel Karol, Jeff Kincaid, Yussuf Lazzouni, Tim Lane, Michael McGuinness, Mike Mullinax, Connor Murphy, Marc Nash, Matt Reilly, Mike Reilly, Matthew Rockman,

Melissa Rockman, Adam Roth, Dan Rowe, Maggie Serra, Daniel Strickland, Gerry Wolfe, Jonathan Ye

TEAM 1319: AdvanceSC, Sealed Air Corp. Cryovac, AssetPoint, National Electrical, Greenville County Schools, and Mauldin High School
Mauldin, SC

Nancy Zende, Chuck Zende, Robert Zende, Erich Zende, Rick Wilson, Hugh Rambo, Glenn Killinger, Jeff Corbett, Paul Way, Crystal Dickerson, Kim Zubik, Catherine Zende, Alex Slessman, Mitch Neiling, Chelsea Slessman, Stephen Corbett, Nick Garrett, Elliot Dickerson, Kate Dickerson, Kathryn Teska, Jarrett Upton, Jon Gawrych, Jonathan Killinger, Michael Hayes

TEAM 1448: Taylor Products, Ewing Marion Kauffman Foundation, Ducommun Aerostructures, Ace Hardware, Ruskins, Dayton Superior, and Parsons High School
Parsons, KS

Matthew Mason, Micheal Bolinger, Zack Strathe, Jacob Breneman, Cole Cooper, Ashton Clemens, Trever Pope, Taylor Rea, Jordan McRay, Bruce Rea, Calvin Schnoebelen, Kim Fentress, Jack Burke, Geoff Sevart

TEAM 1612: State Farm Insurance and Nature Coast Technical High School
Brooksville, FL

Alex Adair, Kevin Adair, Kenneth Alligier, Jake Almasy, Nicholas Anatala, Nathan Barnwell, Hayley Bettenhausen, Natalie Bogicevic, Justin Coleman, Elizabeth Coscia, Justin Fertig, Hank Fyock, Lynwood Gallier, Matthew Granda, Chris Hilbert, Cassidy Holbrook, Aaron Holland, Barrett Karish, Ryan Lowe, Brad Lucier, Nicholas Minjarez, James Raymond, Tamarah Riggs, Emily Russell, Ashley Sedlack, John Sfraga, Michael Smith, Matthew Sosa, Kraig Tatman, Robert Wagner, Michael Williams, Blair Yahn, Oliver Yahn

IMAGES

Each profiled team supplied the images that appear in this book.
Additional image credits:
Adriana M. Groisman, FIRST
Joe Menassa, Joe Menassa Photography